Sept. 1975
To Dave Hughes —
with good wishes to a
good friend!
Ed Feldman

Building Design for Maintainability

*To Hannah and Max Feldman,
my beloved parents*

OTHER BOOKS BY EDWIN B. FELDMAN, P.E.

How to Use Your Time to Get Things Done (Frederick Fell, 1968)
Housekeeping Handbook for Institutions, Business and Industry (Frederick Fell, 1969)

Building Design for Maintainability

Edwin B. Feldman, P.E.

McGRAW-HILL BOOK COMPANY
New York   St. Louis   San Francisco   Auckland   Düsseldorf
Johannesburg   Kuala Lumpur   London   Mexico   Montreal
New Delhi   Panama   Paris   São Paulo   Singapore
Sydney   Tokyo   Toronto

Library of Congress Cataloging in Publication Data

Feldman, Edwin B
 Building design for maintainability.

 Includes index.
 1. Buildings—Maintenance. 2. Maintainability
(Engineering) 3. Architecture—Details. I. Title.
TH3361.F44   690   75-4521
ISBN 0-07-020385-7

Copyright © 1975 by McGraw-Hill, Inc. All rights reserved.
Printed in the United States of America. No part of this
publication may be reproduced, stored in a retrieval system,
or transmitted, in any form or by any means, electronic,
mechanical, photocopying, recording, or otherwise, without
the prior written permission of the publisher.

1234567890   MUBP   784321098765

The editors for this book were Jeremy Robinson and Lester Strong, the designer was Naomi Auerbach, and the production supervisor was George Oechsner. It was set in Caledonia by Bi-Comp, Inc.

It was printed by The Murray Printing Company and bound by The Book Press.

Contents

Preface   vi

1. Opportunities in Maintainability   1
2. Activities during Construction   17
3. From the Ground Up   33
4. Floors, Elevators, and Stairs   62
5. Walls and Ceilings   84
6. Furniture, Fixtures, and Fenestration   105
7. Restrooms, Plumbing, and Piping   134
8. Trades Maintenance   159
9. Maintenance Facilities   179

Suggested Reference Literature   195
Checklist   196
Index   223

Preface

Service Engineering Associates, Inc., is an organization that provides consulting and training service in building maintenance. In the conduct of seminars and schools for maintenance managers, and especially in the custodial field, we have listened to countless complaints about maintenance problems arising from building design that did not take these factors into consideration. A tremendous opportunity exists to make improvements in this field and to save millions upon millions of dollars in the future.

Who should be interested?

- Building superintendents and maintenance supervisors who live with the problem should bring these items to the attention of their management, to be considered both in new construction and in renovation.
- Building owners and managers should stimulate an interest on the part of their architects and interior designers.
- The architects and designers themselves who wish to provide their clients with the greatest possible return on their investment should acquaint themselves with the opportunities in this field.

Certainly, internal operating people, who have experienced the shortcomings of previous designs, should have an opportunity to make recommendations to management, architects, and designers on the basis of preliminary plans and specifications. Naturally, not all of these recommendations would be feasible—because of the initial cost, or certain functional or esthetic demands—but unquestionably there would be some good ideas among them. Further, the use of a maintenance consultant might be considered in such a case.

A word needs to be said at this point about preparations for maintenance from a management and an organizational standpoint. Without an effective maintenance program—either operated by internal personnel or purchased on a contract basis—a "new" building can get "old" in a hurry (here the words indicate condition and appearance). Even many of the effective design considerations recommended in this book will be wholly or partially invalidated if adequate maintenance is not provided. A complete maintenance program can be developed before the building is occupied, and thus a workable program can be installed from the beginning, eliminating errors and mistakes which may be either costly to rectify or which may leave permanent damage. Using the proper techniques, determination can be made of the structure of the building maintenance organization, the number of personnel required, individual job assignments, and equipment and materials required. Programs and controls for both preventive and collective maintenance can be established. Such planning can be looked upon as an investment in protecting the investment in the physical plant that will bring very worthwhile dividends throughout the life of the building.

A work of this type, where individual chapters represent categories of information, unavoidably contains some duplication in order to have each chapter reasonably complete. Cross-references reduce this duplication to some extent.

So much information is constantly being received that plenty of justification exists for unlimited procrastination to delay production of this work. But the problem of maintainability is with us *now*, and action is needed *now*. Thus, this book is a beginning. The help of the reader—*you*—is earnestly sought in correcting and augmenting this beginning on the basis of your own experience.

While claiming full responsibility for any errors or omissions in this work, I also wish to express my heartfelt thanks to the many people who have made the book possible through their suggestions and offerings of information. My special thanks for their time and valuable suggestions go to such friends as Sam F. Brewster, formerly of Brigham Young University, the late Clyde B. Hill of the University of South Florida, John D. L'Hote of the Detroit Public Schools, Clarence P. Lefler of Ohio University, George Moore of the University of Cincinnati, Joseph Novak of the New York Telephone Company, J. McCree Smith, formerly of North Carolina State University, Leland P. Smith of the Cleveland Clinic Hospital, and L. Terry Suber of Colorado State University.

A special note of thanks goes to Margaret McCaffrey Kappa of The Greenbrier Hotel, and to William E. Young of the Sunday School Board of the Southern Baptist Convention, for their vigorous devotion to the concept of maintainability and for their personal example and encouragement.

For a final technical review and editing, I had the great good fortune to be directed by the publisher to Larry C. Dean, AIA, FCSI, of Heery & Heery Architects and Engineers, Atlanta, the then president of the Construction Specifications Institute—and found the opportunity to renew a warm friendship with a former school chum! I am most grateful to him for his many helpful suggestions.

To Jeremy Robinson, Sponsoring Editor of McGraw-Hill Book Company, my sincere thanks for "finding" Larry Dean for me, for recognizing the need for this book, and for encouraging me.

I only regret that all who have helped cannot be named, but space does not permit. Included with these anonymous are several of my colleagues in Service Engineering Associates, whose activities in consulting for our clients have led to some of the best suggestions in this book. And so, unfortunately, these anonymous include some of my best friends!

Edwin B. Feldman

CHAPTER ONE

Opportunities in Maintainability

Between now and the next generation as many new buildings will be constructed as have been constructed in all previous human history! With the great diversity of architectural forms, with the increased attention being given to community planning, and with the greater emphasis that we now see in creativity and expression, unquestionably these buildings will be designed in a profusion of utilitarian and aesthetic designs.

But unless drastic changes are made in the design of buildings from the standpoint of *ease of maintainability,* an enormous opportunity will have been lost; we will be designing buildings in the future, just as we are now, with very little attention being given to the enormous cost of their upkeep.

As if this were not an important enough question now, it will be even more important in the future because of ever-increasing wage rates, fringe benefits, and the difficulty of finding and training personnel for maintenance work in a supertechnological future.

When we speak of ease of maintainability, we mean the condition of an item or a surface that permits its repair, adjustment, or cleaning with reasonable effort and cost.

2 Building Design for Maintainability

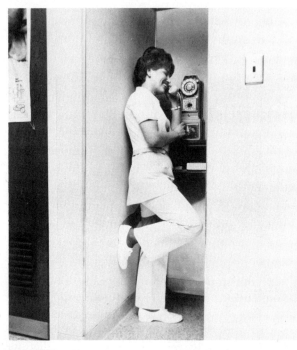

Fig. 1. Flat paints do not serve well under heavy wear conditions; use plastic wall covering or other durable surfaces. (*Rohm & Haas Company.*)

Reasonable effort and cost also means, by inference, that the maintenance must not require unusual worker skills or expensive equipment that is rarely used (although specialized equipment regularly used can be very economical), that it must not involve a procedure that will not permit the reuse of the item in a short time, and that it must not change the item's original appearance or require overly frequent attention.

We can clarify the concept simply, taking housekeeping as an example, by naming a few *nonmaintainable* surfaces (see Fig. 1 for an illustration):

    1. A solid-color carpet, especially of a dark or light color
    2. Low-grade latex wall paint
    3. Soft, blown-on mineral acoustic ceilings (these are more

than nonmaintainable—they simply cannot be cleaned by any means)
  4. Raw concrete block walls
  5. Embossed resilient tile

In order to achieve the best possible results, the combined talents of a number of specialists are drawn on when a new building is designed. We do not wait until construction begins, or has been completed, to consider such things as property drainage, soil loads, climate, electrical distribution, floor loading, equipment layout, sprinkler protection, or transportation access.

Yet why do we complete new structures, or renovate older ones, without considering many of these aspects of ease of maintenance? We fail to consider maintainability so regularly that it has become the rule rather than the exception. In building design, those people responsible for maintenance joke about the possibility of a motto such as "plan ahead—except for maintenance."

But it is no joke. The amount of money being lost each year in the maintenance of our buildings through improper planning and design is astronomical. (See Fig. 2.) For most buildings, over a period of two or three decades maintenance costs will equal the original cost of construction. If only 5 percent of this could be avoided, we would be talking about millions upon millions of dollars. Actually, the potential savings are far beyond the 5 percent figure.

Verify this situation, if you wish, by a discussion with maintenance foremen or managers. Many of them have become resigned to the fact that they will continue to be given buildings to maintain that are simply not susceptible to maintenance—or certainly not to *economic* maintenance. Others are amazed or angered by the fact that, typically, architects, engineers, management, and interior decorators "just don't seem to care." How many of these maintenance managers have bitterly considered the serious problems which could have been avoided if they had been given an opportunity to advise on a simple point! How many of them realize that a number of their difficult maintenance responsibilities are necessary only because of lack of planning!

Many companies and organizations have constructed new facil-

4 Building Design for Maintainability

Fig. 2. Some complicated building interiors can be "blanked out" by design. In this case, the column might have been connected with the wall, or the space used for a storage closet.

ities in order to eliminate the production or service inefficiencies which were unavoidable in the older plants. But, unfortunately, maintenance problems are perpetuated directly from the old facilities into the new—again through lack of planning. Why do we have this apparent lack of interest? Perhaps it is because there has not been enough exposure to the economic importance of the subject; or perhaps too many people do not realize that an opportunity does indeed exist.

The greatest possible return on the maintenance dollar will undoubtedly be realized when the maintenance program is developed during the planning stages of building. Maintenance personnel should work with the architect, engineer, and decorator in selecting materials and surfaces. Working together, they can avoid decisions which eliminate the viewpoint of maintenance—which is equivalent to leaving out one of the variables in an equation.

It is likely that the benefits of maintenance planning will never be fully realized without interest on the part of the owners and managers of the building, whether it be factory, office building, college, hospital, public facility, or any other type of structure. The architect, engineer, maintenance manager, and maintenance consultant all have an obligation to awaken and sustain such an interest. Some architectural firms, although not nearly enough, have already made moves in the direction of obtaining recommendations on maintainability, and even of advising their clients on proper procedures. (See Fig. 3.) A few architects simply will not take on a design job unless management has indicated an interest in good maintenance for past projects.

Fig. 3. Epoxy terrazzo floors are highly durable when properly maintained; the vinyl-covered folding acoustic doors should require limited maintenance; collapsible tables permit ready emptying of an area for maintenance. (*Jack Corgan & Associates, Architects & Engineers, Dallas, Texas.*)

The "newness" of a building can refer either to its date of construction or to its appearance, and thus many "new" buildings have become "old" in just a few short years through improper maintenance, tied in with improper design. In so many cases we seem to end up with less money for maintenance than was at first felt desirable, but with proper design even that lesser amount of money might well have been sufficient to keep the building "looking new." The opportunity to properly maintain a new or renovated facility from the first day of use comes but once, and once lost is gone forever. Thus, planning for new construction or renovation should also involve planning for sound maintenance staffing and operation, thus assuring that the facility will stay in "new" condition for many years.

A TOUR OF INSPECTION

To better view the opportunities that exist, let us take a walk through an imaginary building (a building that may be all too real to you because of the problems you have to live with every day!). Let us keep our eyes open for examples of poor maintainability. It will help us develop a viewpoint so that, hopefully, the problems we come across can be avoided in the next construction or renovation. For simplification, let us assume we are on an inspection tour for custodial conditions.

Our tour begins before we enter the building. Almost the entire world can be considered as soil, as far as our building is concerned, and we want to keep as much of that soil out as we can. How much easier it is to keep a pound of dirt out by having it drop into the catch pan of a grating than it is to spend perhaps $100 removing it later from floor and furniture surfaces. The perfect place for the catch pan is right outside, immediately in front of the entrance door. The pan is emptied by sliding it out sideways, so that the grating does not have to be lifted. It also provides a "weep" system for the runoff water.

Another way to keep out dirt, sand, mud, and water is to use grid carpets. Individual carpet treads, which can be replaced economically when desired, come in blends of long-wearing synthetic fiber.

Also useful, and maintenance cost cutting, are mechanical floor mats. These devices are used at lobby entrances to remove dirt, moisture, and soil from the soles of shoes. The machines cut maintenance costs, eliminate slippery floors, increase the life of floor coverings, and reduce contamination.

Even the revolving door can have a circular grating unit.

In addition, the walkway should be crowned, so that puddles and soil collections are not created. In some climates, ice-melting systems should be considered.

For a vestibule, or perhaps for the lobby itself in the entrance area, a rough granite surface will collect soils before they can be tracked to carpeted, terrazzo, or resilient areas beyond.

An ideal soil-entrapment system would use a combination of devices, perhaps in this order: crowned walkway, snow melter, grating, matting (preferably recessed) and/or runner (see Fig. 4), rough stone, and carpet. Yes, carpet is a good soil trap, and sometimes a sacrificial square of carpet, to be rotated at

Fig. 4. Although the matting is useful, a grating with catch pan beneath, followed by a recessed matting, would be both more effective and safer.

intervals to avoid traffic patterns forming too quickly, is a good investment. Elevators and stair landings especially should be carpeted to avoid soil transference from one floor to another (and also because it is so much easier to clean those areas than a whole floor).

The subject of carpets has been thrashed out so exhaustively that this short work is not the place for a full treatment. But perhaps we can agree that, if properly selected, there are areas where carpet is a good, maintainable surface. The wrong carpet, or the right carpet in the wrong place, would not be. At least try to avoid deep, cut pile and use mixed colors—especially four-color tweeds.

Perhaps our imaginary building has a number of isolated executive offices that are carpeted (a mark of status), while surrounding areas are resilient. Or the corridor is waxed while the side offices are carpeted. I have heard this condition called "islands of misery," because it is very hard to maintain such mixed areas. Think only of the different equipment, training, and systems needed as well as the possibility of damage to the carpet while stripping an adjacent waxed floor. Would it not be better to carpet the entire area?

In our tour, we come across a vinyl asbestos floor that has swirling, concentric ridges in its surfaces. Here we realize that the cleaning problem (which can never be solved with this floor, short of replacing it or covering it with carpet) is not one of design, but one of installation. The flooring contractor used a serrated trowel in applying the mastic, and applied it in a too-stiff mixture, let it harden before applying the tiles, or did not roll out the floor—or some combination of these things.

In the hands of a skilled worker, by the way, a paint-type roller applicator does a much better job of mastic application than a serrated trowel. A poor worker can leave spaces between tiles or use too much mastic, which squeezes up between the tiles and remains a problem for months or years. In any event a floor that is not right should not be accepted from the contractor. It might take a month for foot traffic to reveal a "problem of the ridges," but the construction contract should provide for rejection for this cause even at that stage. The best solution

Fig. 5. The correct use of the dark grout in the floor tile provides a uniform appearance; the same treatment should have been given to the wall.

is to use a capable flooring contractor and to check his work during application.

We are about to pass a restroom. Let us look in. The first thing we notice is the ceramic tile floor (people generally look down when walking, and floors are typically the first thing noticed). The floor has a white cement grout, which was a definite mistake in planning because it shows all soils. A tan or gray grout is preferable by far. (See Fig. 5.) Of course, whatever color it is, it should have been sealed before soils were allowed to penetrate such a porous material.

The partitions are floor-mounted rather than ceiling hung, which creates quite a problem in maintenance. Similarly, the fixtures themselves should have been wall-mounted to avoid problems in floor care. Perhaps the room is inadequately ventilated to remove transient odors. Sometimes correction can be made by using vent covers that expose a larger percentage of area, or by putting grills in the door. Single dispensers for tissue and towels should all be replaced with double dispensers, to eliminate complaints and reduce tending time. In the men's toilet room, metal partitions should be completely eliminated between urinals as they quickly become a maintenance problem. Selection of urinals with built-in side panels of ceramic material are preferable.

It is now convenient to look at a typical custodial closet, since it is economically practical to locate it next to plumbing fixtures. The most common mistake with custodial facilities is inadequate space. The typical custodial closet should be about 6×8 to 6×10 ft—large enough to store the equipment and materials needed daily by at least two workers. This saves a great deal of time that would otherwise be wasted by the workers in moving back and forth between the work area and a central supply point. Let us also remember that if workers have to leave their assigned area to obtain equipment and supplies, we have given them a built-in excuse to be absent from their area at any time they choose.

On the other hand, it can be a mistake to provide too large a storage area. We have seen many cases where custodial closets of 100 ft^2 or more became so attractive to other departments that they were soon taken over by them, and now the housekeeping department has nothing but what remains under a stairwell (and this would probably be in violation of the fire code!).

To be considered along with the custodial closet are other areas used by the housekeeping department, such as its central storage area and its own offices. Normally, a set of three office areas is required, one for the executive housekeeper (with sound conditioning to permit complete privacy during conversations and interviews), one for assistant housekeepers, and one for clerical or secretarial help. These should be next to the central storage area, which should have room not only for supplies and equipment, but also for cleaning and repairing these items. Here might be located a washing machine, solution tank, steam cleaner, and the like. For large departments with considerable turnover, a training area would also be called for.

Perhaps now we begin to notice the windows in this building we are touring. Maybe they are the pivoting type, where, with the turn of a key, the window can be rotated (preferably on a vertical axis) to bring the outside in, so that it can be washed in any weather or at any time of day by other than high-paid specialists. For rotating windows to be practical, the building must be on a firm foundation, as settling may cause the windows to bind. Although the theory of the "sandwich type" of pivoting window, with a Venetian blind between two panes of glass, is excellent, these designs have not been perfected to the point where they are uniformly successful. Hopefully, this will come soon.

Sad to relate, we have seen excellent pivoting windows installed in buildings, only to have valances (put in by a decorator) located so that they came down below the window top, thus making it impossible to pivot the windows after all. In other cases, we have seen fixed window glass covered with decorative masonry that left a space between them of only about a foot, making it literally impossible to clean the outside of the glass.

We now come across a wing of the building that has been placed at an angle to the main building. This would create an equal number of obtuse angles (which are no particular problem except for furniture placement) and acute angles. It is these latter cases that create a problem in terms of painting and cleaning. Any floor cleaning in a sharp angle, for example, requires hand work and usually ends up in a blackened condition after a year or so.

It is at this point that we might have completed our tour, observing and considering just some of the many factors that are involved in custodial operations, which in turn is one of the several types of maintenance. Perhaps the tour has left us unhappy with what we have seen, and determined that these mistakes will not be repeated in our next construction.

This determination can best be served by recognizing the opportunities and "selling" them to interested parties.

DESIGN COMMITTEE

From the inception of a construction project, various forces tend to weaken maintenance design activities. In many cases, inflation creates such a demand for additional funds that compromises may be required in maintainability. Delays due to strikes, weather, design changes, or other causes compound the money drain and increase interest costs. Our rapidly changing technology can also require modifications, such as for computer installations, that can be very expensive. Another factor undermining maintenance planning is the all-too-frequent rush construction job. All these forces also tend to lead to inadequate operating funds for maintenance, which is doubly bad. The answer to these problems falls in two categories: First, many of the problems can be anticipated so that proper planning can be done, with outside help, if necessary. Second, consideration should certainly be given to the proposition that it may very well be more advantageous to have, say, a 10 percent smaller building with a 25 percent lower maintenance cost because of correct initial investment in maintainability rather than a full-size building that will never be properly maintained because of lack of funds. Where the rules of maintainability are broken, a new building can age a year in a month's time.

Persons who design buildings for construction often talk a different language, so to speak, from those who will later operate those same buildings. In some cases, for example, the architect may be no more able to appreciate the problems of a maintenance manager than the maintenance manager is able to appreciate the structural and aesthetic considerations of the architect. It is management which stands squarely in the middle between these two viewpoints, and it is management which must bring these viewpoints into a concerted action which will provide the optimum balance between construction and operating costs. This may best be done by requiring frequent meetings between the two parties, with management acting as moderator. Speaking for management should be an appointed coordinator who is aware of both construction and operating matters, but who is a practitioner of neither. Coordinators retained by management would typically serve management directly in a staff category, rather than serve either the design or the operating viewpoint.

You may wish to consider the formation of a committee to make recommendations concerning maintainability. The committee would make recommendations (always, of course, carrying both initial and long-term cost figures) on which management would have the final say. It is quite likely that the committee's suggestions would not always be unanimous but would be supplemented by minority viewpoints. Such a committee might include representatives of the architect, interior decorator, custodial manager, mechanical maintenance manager, facilities engineer or building manager, and perhaps a consultant. Coordination and action on recommendations should be the function of an assigned member of management, possibly in a staff position to a chief executive officer.

A worthwhile preliminary to the first meeting of the committee would be the requirement that all members be prepared with checklists which they have developed on the basis of their own experiences and of the experiences of those with whom they are in contact. Where a recommendation appears on more than one list, it may be possible to reach a speedy consensus of its desirability; nevertheless, some very worthwhile actions may appear on only one list or perhaps initially even on none at all.

In the development of the checklists, and in the subsequent recommendations, some cases will appear where more than one action is possible; in these cases, choices should of course be made on the basis of desirability.

Following the completion of the construction project, a "postmortem" should be held by the same committee approximately one year after occupancy so that a report can be given concerning the results of its actions, both those recommendations acted on and those rejected. This follow-up report might become the basis for a companywide file on maintenance design. Perhaps an individual would be appointed to keep the file updated, and to make periodic reviews—such as every few years—of the effects of building maintenance design.

ECONOMIC ANALYSIS

A final selection of surfaces, fixtures, and other recommended items can only be made by first limiting the number of possibilities to be considered on the basis of alternates acceptable to management from an aesthetic standpoint, coupled with an economic analysis of the acceptable choices. The analysis must include an estimate of the probable life of the item, the annual cost of maintaining it at the quality level desired by management, as well as the initial cost of purchasing and installing it. The variables involved in estimating these factors, along with local wage rates and good prices, maintenance cost, and manufactured quality are so complex as to require a cost analysis in each case. Once such an analysis has been made, however, it may be used regularly thereafter, with minor adjustments for wage and price changes.

These are the factors involved in a complete economic analysis:

1. Initial cost
2. Anticipated man-hour savings
3. Anticipated wage rates
4. Anticipated cost of fringe benefits and support costs
5. Comparative life expectancies
6. Cost of money (interest costs)
7. Salvage value
8. Effect on insurance, if any
9. Effect on taxes, if any
10. Effect on depreciation, if any
11. Secondary benefits

It is unavoidable that, at times, selection will be made strictly on the basis of an architectural or aesthetic effect, without regard to economics. In such cases, management at least deserves to know the full implication of the decision it has made through an economic analysis of the annual cost of such a decision.

Maintenance departments are typically involved in the following activities, which should be included in cost-improvement planning:

1. Maintenance inspection
2. Preventive maintenance activities, such as lubrication
3. Corrective maintenance, such as repair or replacement of broken or worn items

4. Cleaning and sanitation
5. Renovation and alteration (closely allied to maintenance)
6. Utilities control (also closely allied)

When consideration is given to making initial investments in construction so as to minimize future maintenance expense, the natural conflict between initial outlay versus repeating future outlay is often heightened by the effect of rising construction costs. The forces of inflation or the interest costs imposed by delays may require changes in physical planning that can have a serious adverse effect on future operating expenses. Changes may be unavoidable with respect to preserving the financial integrity of the venture; but even in selecting less expensive alternatives, the same principles of *design for maintainability* elsewhere expressed in this book should, and can, be adhered to. Remember, it is often not realized that a number of changes *beneficial* to maintenance could be incorporated in the planning, at no more cost or even at less.

Building owners have a real stake in ecology, too, since a polluted atmosphere can cause heavy soiling of building exteriors and can actually corrode metal surfaces, leading to expensive cleaning, painting, sandblasting, replating, and replacement costs The building owner has a very practical interest in cleaner air!

Inadequate cost planning for new buildings often results in a "tightening of the belt" in the later stages, which more often than not means making cuts in expenditures for internal finishes. And since the cost of these internal finishes themselves is such a small part of the total building cost, any reduction here has a marked or even drastic effect on performance and maintainability. Naturally, this situation is more prevalent in times of inflation than otherwise, since it causes the money to run out before the building is complete. This particular problem may be considerably alleviated with some of the newer forms of rapid, coordinated design-construction. Finishes and other relatively "minor" consideration should probably be given full attention as early as the type of design program permits, and once the costs are fixed within the total building concept, they should be considered no more subject to change than earlier expenditures for land, architectural fees, and foundations.

Think of all the buildings that will be designed and constructed during your working career—what monuments to economic waste they will be if they are designed with the same lack of regard for their maintenance as we have seen in the past!

CHAPTER TWO

Activities During Construction

If at all possible, the maintenance manager should be on the payroll during the latter stages of construction of the facility, if it is other than the simplest sort of building, such as a prefabricated metal warehouse. These personnel should make regular inspections of the facilities during construction so that they will be aware of the materials involved, the location of items that will later be hidden from view, and the easiest means of access. At this time they should also take photographs and prepare diagrams and plans to supplement those which would ordinarily be provided. A secondary benefit of such inspections is the possibility of *observing errors of construction* which might be called to the attention of the architect and corrected before it is too late.

A still further step can involve a full-time inspector hired for surveillance during construction. Organizations which are continuously constructing facilities—such as some industries, universities, public school systems, government—may have one or more long-term personnel of this sort. If these people have operating maintenance experience, they can provide an invaluable contribution to the later maintenance of the building; if they do not have such experience, their inspections should be supplemented by operating personnel. Naturally, these inspectors in no way minimize the importance of inspections made by the architect's staff, but their background and orientation is different so that the groups complement each other.

So many new materials and so many new items of operating equipment are coming on the market that it is often tempting to take a chance on an innovation that has not been properly tested or approved. It is not that new systems should never by used; rather, they should be well tested and proven before being installed on a large scale. Where an item looks particularly desirable, a test run may be provided in some existing facility. Reckless experimentation can lead to extremely high maintenance costs and when it comes to systems installation, remember that our ever-increasing sophistication requires ever-increasing maintenance requirements. As one becomes more complex, so does the other. Unless the owner is prepared to provide sophisticated maintenance, then simpler installations should be used.

Standardization is an important part of maintenance planning; the greater the variety of items—such as valves—the greater the problem in training personnel for maintenance, stocking or obtaining of replacement parts, and even of identification. Of course, standardization can be overdone, leading to the continued use of an item that has long been superseded. Any standardization list must be reviewed and updated regularly. In general, however, standard items should be specified unless the exception is justified. In other words, the burden of proof is on the person who wishes to break the standard.

THE CONSTRUCTION CONTRACTOR

The role of construction contractors, although far less important than that of architects in determining maintenance costs after construction, is still significant. Their activities will have a considerable effect on costs, some of these lasting for the life of the structure.

If the contractor is approached before the work begins, it is possible to come to an agreement on construction methods which will considerably reduce the amount of dirt and dust created. Such methods can actually be written into the construction specifications, and in many cases will not increase cost appreciably. Where cost is increased, it must be considered in the light of its value as preventive rather than corrective action. Some of these methods, or steps, include the following:

1. Specify wet grinding rather than dry grinding for a number of operations, as this will sharply limit the dust problem.

2. Instruct laborers to clean up accumulations of dirt and dust and remove them before they are spread.

3. Remove waste, particularly plaster and concrete, in such a way as to avoid clouds of dust. This means avoiding dumping materials from upper levels (usually a waste chute is used). Keeping such waste wetted down also helps.

4. Use "duck boards" on rainy days to prevent mud from being tracked into the building.

Typically, it is the responsibility of the construction contractor to provide a preoccupancy cleaning of the constructed or renovated areas so that they are ready for actual use. To many building owners, the quality of this cleaning, and the types of surface materials—such as seals and waxes—that are provided by contractors before final delivery, are not acceptable, and the owners often require complete removal of these finishes and replacement with high-quality materials. One of the most common faults of construction contractors is their failure to thoroughly clean floors before they apply a finish to them, thereby sealing in dirt, grit, and other forms of soil.

Great care should be taken in specifying preoccupancy cleaning to be provided by the contractor, perhaps with the actual cleaning materials—especially the floor finishes and sealers—being provided by the owner (and their value being reflected in the contract price). Although the theory of removing preoccupancy cleaning from the construction contractor's requirements and hiring a contract cleaning organization or using one's own custodial staff to do this work is good, it may lead in practice to the contractor taking an attitude of indifference toward the floors, since he knows that the preoccupancy maintenance will not be his responsibility. The quality of workmanship and activities of the contractor throughout the construction will be important for a great many items, from a maintenance standpoint, but below are discussed some of those that are most significant. Perhaps this will serve as a checklist for developing your next specification and inspecting the work.

Especially in a renovation or addition project, but also in a complete new construction, the contractor should be required to use a type of heat that will not cause stain and soot to be deposited on surfaces.

The contractor should be required to use building paper to protect final finished floors. The paper should be used copiously and held in place with masking tape or similar means where necessary. Where the paper is torn or worn through, it should be replaced or doubled over. Where there is exceptionally heavy wear, carpet runners or old discarded carpeting can be used in place of the building paper.

TRADES WORK

The condition of the concrete surface on which resilient tiles will be placed—and even on which carpet will be placed, under certain conditions—will have a telling effect on the final surface. Marks left by manual troweling will show up as a permanent ridge in the surface, which will cause both unsightly appearance and rapid wear of the resilient material or carpet. Steel troweling by machine can do much to help in overcoming problems of this type, but the job must be performed by skilled workers. (See Fig. 6.)

Improper application of resilient floors, such as vinyl asbestos or rubber tile, can lead to serious maintenance problems for many years after occupancy. A common method of applying the black mastic which holds the tile in place is with a serrated trowel. If the mastic is applied when it is too stiff, or if it is allowed to harden somewhat before the tile is applied, and especially if a weighted roller is not used to flatten out the mastic under the tile, then the tile will tend to stay on top of these mastic beads. Hardening there, the tile will then tend to sag down between the beads, taking the shape of the swirl patterns which the beads themselves took in being applied with the

Activities During Construction 21

Fig. 6. A properly applied resilient tile floor, over a well-troweled subfloor, allows for rapid cleaning with an automatic scrubbing machine. (*Advance Floor Machine Company.*)

trowel. Thus, we end up with a floor with a number of swirling concentric grooves, and no amount of waxing or other surface care will eliminate this most unsightly condition. It may be possible, using heat and pressure, to soften the mastic underneath, but this is a hazardous procedure that may end up in a completely damaged floor. Contrarily, the mastic may be applied in too fluid a condition, or in temperatures too low—or simply too much of it may be put down—so that it later comes up between the joints in the tile. This migration of the mastic continues for months or years, and the sticky black material is tracked around from place to place. The solution to these problems is to demand tile installation by a qualified person only, and to urge that the mastic be applied with a roller rather than a trowel; floors which have been applied with too stiff mastic or mastic which comes up between the tiles should be rejected; and it should be borne in mind that these conditions may not become apparent for thirty days.

Grinding of terrazzo or concrete should be done with a vacuum attachment device, thus creating far less soiling than without such a device or even by wet grinding.

Disposable, temporary filters should be used in the air conditioning system during construction to prevent clogging or plugging of permanent filters.

The most common problem in the installation of restroom fixtures is mounting urinals too high, the end result of which is considerable floor staining. Attention should also be given to the placing of other fixtures so that the valves and flushing mechanisms can be repaired or replaced without wall damage.

There is a big joke in the building maintenance industry that there is an unwritten rule that all floor drains are to be installed at the highest point of the floor. Of course, the humor disappears when this actually happens, which seems to be quite regularly. It is certainly worth the extra time and effort required to assure that all floor drains do indeed end up at the lowest part of the floor.

In painting operations, make sure wood sash and screens are free so they are not painted closed.

Any grease-pencil marks or other foreign materials on walls must be completely removed before the application of vinyl wall coverings to prevent bleeding through later on. Joints must be carefully matched and cemented.

Contractors not infrequently leave ceiling tile in an unsightly condition, most often marked with hand prints but sometimes also marked with guidelines and other instructional material for installers. No ceiling should be accepted which is in any way marked, since such surfaces are extremely difficult, if not impossible, to clean. All marked tile should be replaced with new tile before acceptance.

EXTERIOR CONSIDERATIONS

No topsoil should be buried in fill, as it is a material of vital importance in grounds care later on. The topsoil stripped from the construction site should be stockpiled, and at the completion of the job it should be spread on the ground areas around the

Fig. 7. Care should be taken to eliminate unsuitable materials from ground fill in order to eliminate slumping and building settling.

building, following scarifying 1 ft deep to remove all construction debris (Fig. 7). Particular emphasis should be placed on the selection of backfill material around the walls of a building:

1. Ensure the removal of mortar spoil or other debris which is injurious to plant growth.

2. Remove objects such as broken blocks or stones to prevent puncturing of waterproof membranes.

3. Remove wood, paper, and other such material to prevent fostering of termite colonies and later subsidence of the backfill.

4. Place backfill in layers not more than 6 in deep, and thoroughly tamped.

Improperly compacted soil can have a most far-reaching effect on many aspects of building maintenance. Such a condition is nothing less, in some cases, than a disaster, since it results in such things as badly cracked terrazzo floors, jammed doors, pivoting windows that will not pivot, cracked walls, stuck valves, sprinkler systems that do not work, and so on. Soil compaction, although at first sight far removed from building maintenance, is in reality a vital consideration in this subject. One unfortunate thing about this problem is that it may not show up for many months—long after the contractor has been paid in full.

Grounds require special attention, both immediately before and during construction.

If a lawn, flowers, ivy bed, or the like is to be saved, the protected area should be staked out and roped off, at about a 3-ft height. If necessary, cloth or plastic streamers can be hung from the rope to improve visibility.

If walkways or other paving are to be left intact, the thickness should be investigated, as the type of construction equipment used, or delivery vehicles expected, may damage the surface. It may be necessary to reinforce the paving with steel plates, to provide temporary entrance areas to avoid the older paving, or simply to sacrifice another area of grounds which would later be repaired.

Should any mud or debris collect on the streets or walks from the construction project, it should be removed immediately before it becomes a traffic hazard or is carried into the surrounding buildings, either occupied or under construction.

All catch basins and storm-drain lines near the construction site should be protected at all times from the entry of mortar, concrete spoil, and other construction debris.

Trees and shrubs designated for protection during the construction program should be properly marked by the owner (typically with a colorful plastic ribbon or band). This protection should be provided before any mechanized equipment is permitted on the job site. It should consist of barricades composed of 4-in square posts which are driven at least 3 ft into the ground, the posts connected by 2×8-in planks. Further, 4-in-square timbers securely wired around the trunk of the tree should extend from the ground to the bottom branches, or at least 8 ft high.

Protection should be given to surfaces which are susceptible to damage during exterior cleaning, such as aluminum, stainless steel, chrome, and white brick. Muriatic acid, for example, can damage all these materials. When chemical cleaning is used, care must be taken to prevent etching of the mortar joints or damaging of other materials. All chemical residue should be thoroughly cleaned from the masonry surface. If sandblasting is permitted, special procedures should be followed to assure that other surfaces are not damaged, and that the surface being cleaned is not etched or the mortar joints damaged.

The architect should indicate elevations on floor drains, and the floor should be sloped to these drains at a minimum of $\frac{1}{4}$ in per lineal foot. Specifications should indicate that if floor drains are not installed on that basis, the contractor will be required to lower the drain and be responsible for the repair and restoration of all other work damaged by this correction. Roof drains should be handled similarly. Where economic and structural conditions dictate that ceramic tile be installed by the thin-set method, a 12-in dished area around the drain is very helpful.

CUSTODIAL MAINTENANCE

Custodial maintenance is of special concern during building construction and renovation. Even where the building is not related to any adjacent facility, loose debris, fasteners, and other items must be kept off the floors, as they can cause permanent damage. A specific example is the screw or bolt that is all too frequently dropped into a terrazzo floor, later to show up in a cross section when the floor is ground—or a piece of electric wire which is forced into a vinyl asbestos floor under the pressure of the wheels of a cart, thus making a permanent snakelike groove in the floor.

A special and serious problem arises, however, when the new construction is immediately adjacent to an older facility which remains in use, or where the work is the renovation of an existing facility. Unquestionably, such construction and remodeling create problem conditions which require intensification of maintenance activities in the adjacent areas.

Interestingly, building managers often assume that an area which is being renovated and had been normally maintained by, say, two custodians, permits the reduction of the cleaning staff by the same two personnel, since the area is not in use. In practice, this is typically not the case. The maintenance requirements for the balance of the structure, caused by the creation of various types of soils during construction, intensify the problems in those existing areas, so that often not only is it necessary to retain the entire staff, but even to augment it with part-time workers, cleaning contracts, or with temporary employees. The problem is somewhat similar to the extra effort required in normal building maintenance because of inclement weather; it becomes another source of unusually heavy entry of soils of various types. Failure to consider such needs leads to problems, some of which are permanent and some temporary. For example:

1. As mentioned, permanent floor damage can be caused by scratching or abrading through the tracking-in of particles of concrete dust, metal chips, sand, splinters, mud, wire fragments, and the like. Not only is this true for resilient flooring, but harder floors such as marble and terrazzo, and even stone, may also be damaged.

2. Dry soils which are permitted to accumulate on floors quickly become airborne, and this dust (which may contain fine metal particles from grinding or welding) can damage door hinges, machine bearings, electric motors, data processing equipment, and office machines.

3. The state of mind of employees, visitors, and the public is not a factor to be treated lightly. People are becoming free in their criticisms of unsanitary and even unhealthful conditions caused by construction soils; the next step is the involvement of the federal government in terms of safety and health protection for the employees. This consideration is even more acute in terms of health-care facilities, such as nursing homes and hospitals, because of the recognized relationship between soils

and cross infection, since disease germs are often transmitted on particles of dust.

In the handling of construction soils, the principle that should control most activities is the isolation of the construction area from other areas that are in use, like isolating a sick person from a well person—and the area under construction, in this case, can certainly be considered "contagious"! Fortunately there are a number of ways available to effect this isolation:

1. Provide "air locks" in hallways which will tend to confine dust to the part of the building under construction. These should be made of two sets of fabric, canvas or plastic curtains, which must be pushed aside in order to walk through and yet will fall back into place. They should be the full length and width of the hallway opening. These curtains should be vacuumed regularly—at least once a week. From a safety standpoint, it is desirable to have at least one portion transparent. Be sure the area between the curtains is well lighted.

2. In renovation projects, where aesthetic and design considerations permit, cover ornate ceilings and eliminate drop lighting fixtures by using a false ceiling; old room heights usually will accommodate this.

3. Place plenty of mats and runners in connecting areas to entrap as much dirt and soil as possible and to keep it from being ground into flooring and tracked through the building. The ideal location for these is within the "air lock" area. Mats should be vacuumed daily and washed at least once a week. The use of old carpeting is very helpful; some organizations save discarded carpet for this purpose.

4. Check the filters in the forced-air system regularly so they do not become "loaded" with construction dust. An efficient filter system will cut down on airborne dust.

5. Put up "off limits" signs to prevent construction workers from tracking debris into areas not involved in their work.

6. Put up signs to prevent sightseers from walking into the construction area and tracking soil back into the original facility. To better enforce this rule and at the same time satisfy a natural interest, supervised tours of the new areas might be conducted.

7. If any of the owner's equipment or machinery located near the construction area is not regularly used, cover it with plastic sheeting.

The above techniques will not completely exclude all soil from the occupied area, although they will greatly minimize it. Nevertheless, the remaining soil in the occupied areas is a serious consideration, and these steps should therefore be taken:

1. Floors must be wet-cleaned more regularly to remove the heavier particles which become embedded in the surface and its protective coating. Daily use of autoscrubbing equipment is particularly desirable in open areas.

2. Walls, trim, and furniture should be vacuumed periodically to prevent formation of permanent soils which occur when dust is permitted to unite with water or other vapors and oils.

3. Once construction is complete, a thorough washing of walls and trim may be needed.

4. A protective coating of floor wax, finish, or seal should be applied to all floors before construction actually begins in order to confine as much wear as possible to the protecting material. Floor coating should be stripped and reapplied much more frequently than under normal conditions to prevent hard materials from being ground through the coating down into the floor. When the project is complete, a terminal stripping and rewaxing of all affected floors is desirable.

5. Connecting areas between the existing facility and that under construction should be given especially careful attention, with daily wet cleaning the rule.

6. To prevent transfer of soils from one floor to another, plan more frequent cleaning of stairwells, elevators, and entrance areas. Carpeting is of particular value in elevators and entrance areas.

All this extra work requires more manpower—more than the existing custodial department could handle other than for a short time. The additional man-hours might be provided in one or more of these ways:

1. Present employees may be used overtime. This is undesirable except for a short time, because persons who become accustomed to overtime feel deprived when it is discontinued and a

morale problem results. Also, the fatigue factor and consequent reduction in efficiency must be considered.

2. Temporary employees may be hired, either specifically for the extra jobs required or else to permit a redistribution of the work.

3. Additional full-time permanent employees might be hired to handle the extra workload and then become the cadre for the staffing of the new facility.

4. Contract cleaning firms can be engaged to perform specific jobs at assigned frequencies, or to clean entire areas for a fixed period. It is important to specify materials, methods, frequencies, and supervision when using such services.

Although these activities and precautions sound involved and troublesome, they are nevertheless by far the lesser of two evils. Proper maintenance is a short-term activity during construction and comes to an end within a matter of months or a couple of years at the most; failure to provide these services leads to problems with a duration limited only by how long the building will last.

FINAL ACTIVITIES

The architect normally delivers to the owner, before final acceptance of the job, a corrected and completed set of as-built drawings. Periodic checkups during construction may be necessary to be sure the contractor is providing information to the architect so that drawings can be corrected to the as-built condition. A complete set of these as-built drawings (construction drawings) should be mounted with all edges bound, in a metal cabinet for this purpose. It should be made clear that no one is to remove these master plans from their permanent location. In addition, a microfilm record should be made of the drawings and kept in a fireproof location. Similarly, all instruction manuals, guarantees, certificates, and permits should be properly protected and kept in a permanent location.

Before the final acceptance of the work, the architect should provide the owner with three complete sets of all operating and maintenance instructions. These instructions should be

bound in a quality grade hardback binder and should contain full details for the care and maintenance of all visible surfaces and all mechanical and electrical equipment. The following information should be included:
1. Complete description of items including their catalog numbers
2. Complete parts list for each item
3. Name and address of local supplier
4. Name and address of manufacturer
5. Complete operating instructions
6. Complete maintenance instructions

One of the most beneficial steps that can be taken to optimize maintenance costs before the occupancy of a building is the development of a complete maintenance operations manual. Based on analysis of square footage, surface types, area utilization, fixtures, furniture, and equipment, and coordinated with management's quality/cost objectives in maintenance, a complete staffing and individual job assignment can be determined in advance based on maintenance time standards, and taking into consideration additional manpower elements for such items as relief of absenteeism, emergency work, special services to tenants, special services to other departments, and the like. Based on such an analysis, not only can budgetary requirements be established in advance for supervision, operative and support manpower, maintenance equipment, supplies, facilities, and the like, but also a great many mistakes can be avoided by "doing it right" from the beginning. Where maintenance programs are established other than on a planned basis, errors can be made that can lead to considerable expense because of damage to equipment, marred surfaces, breakage, and other problems. A preplanned program is also desirable from the point of view of the work force who, under other conditions, undergo the anxieties of trial-and-error or "guinea pig" arrangements.

It should be stated in the specifications that the contractor shall keep the premises free from an accumulation of waste material, rubbish, and garbage at all times. Tramp metals, such as stray screws or nails, are a special problem that should receive special attention, as they may cause considerable damage to resilient floors, terrazzo, stairs, and may even find their way into mechanical equipment. On completion of the work the contractor should remove all rubbish from within and around the building. Stains from paint, tar, mud, and other such substances should quickly be removed from porous materials such as concrete, grout joints, and so forth, before they have an opportunity to penetrate too deeply into the surface.

The improper balancing of hydraulic and ventilation systems by the contractor results in the expenditure of what appears to be an endless amount of time on the part of the maintenance staff. All such balancing should be performed in the presence of a representative of the building owner, and the balance records should be provided to the building maintenance department. The contractor should use the services of an independent test and balance agency specializing in the testing and balancing of air and hydronics systems. Members of either the Associated Air Balance Council (AABC) or the National Environmental Balancing Bureau (NEBB) are acceptable, but they should not be affiliated with contractors who work on the job. Proper system balancing, by the way, is extremely difficult if not impossible during "beneficial occupancy" (see below).

Before any building is accepted from a contractor by the owner, all systems should be tested and demonstrated. This should include such frequently overlooked items as:

    1. Roof drains clear and draining properly

    2. Floor drains in bathrooms, janitor closets, and other areas clear and draining properly

    3. Sprinkler system charged

    4. Fire alarm and smoke detector functioning

    5. Lawn sprinkler system operational

It is not unusual for owners to want to move into parts of the building before it is fully completed—this is known as "beneficial occupancy." Although this appears "beneficial" to the owners because they can use the facility before it would normally be available, in practice it can cause many difficulties. Contractors are relieved of many responsibilities through this action, and they may also use it as an excuse to delay completion while perhaps working on other projects. Beneficial occupancy may require a written release from some of the responsibilities on the part of the contractor; this should be avoided whenever possible.

A transitional aspect of planning occurs at the end of construction when the owner is actually moving equipment, furniture, supplies, and other items into the structure, in many cases before it is complete. Such moving should be carefully planned in order to avoid damage to walls, doors, and other surfaces that would require corrective maintenance or replacement. For a large operation, it might be desirable to consider specialists in the field of *planning* the move, as well as in actually *making* the move. Specialized equipment for moving furniture, such as desk and file movers, is often a good investment. Consideration should be given to renting equipment for a one-time use, or purchasing it if regular internal moves are expected.

Finally, no new or renovated facility should be used without a promotional device aimed at securing the cooperation of the users of the facility toward better maintenance. The responsibilities of the individual should be outlined in terms of preventing damage to surfaces; avoiding undue litter, soil, and spillage; defacing property; mounting posters, calendars, and signs; and related items. Such material should be made a part of an employee handbook or guide.

CHAPTER THREE

From the Ground Up

Much of what later happens on the *inside* of a building—and keeps on happening for decades, if not generations—is directly affected by what decisions are made concerning the *outside* of that building. Considering the design of buildings from the standpoint of maintainability, it is appropriate to start from the ground up—or rather from *below* the ground up. This chapter, then, is devoted to maintenance-design subjects outside a building, including grounds, outside building surfaces, and roof.

BELOW GROUND

The initial treatment of the building site for termite control, although it is a first cost which can obviously be avoided, could save a great deal of expense in later years, especially in areas which are susceptible to termite damage.

LANDSCAPING

Rising now to the surface of the ground, we enter that realm of growing things that provides us so much beauty and pleasure on the one hand, and so much cost and difficulty on the other. Some people feel that landscaping gives a building, or group of buildings, a distinctive character; they even carry this point further by stating that those buildings without landscaping have no character!

Lawns are often the fundamental aspect of landscaping, the matrix in which all other features are fixed; and in many cases lawns are the greatest investment in landscaping, both initially and in their upkeep. Our first consideration, simply, is to provide ground on which the grass can grow without undue investment of gardening time. Here are a few guidelines:

1. Provide soil suitable for growth; attempts to develop an attractive lawn on soil which can not support it (chemically, materially, or organically) can lead to a great waste of money.

2. Rather than trying to grow grass in difficult locations, such as on steep or rocky slopes, or in bumpy or rough areas, use low-maintenance ground cover. Similarly, avoid the large amounts of time necessary to grow grass in difficult places, such as where there is a great deal of foot traffic, under roof overhangs, or in heavily shaded areas.

3. Remember that ideally grassed areas should have a 2 percent slope. A lesser degree causes puddling and makes grass growing difficult; a greater degree gives problems in soil erosion. Of course, for proper drainage the slope should be away from the building. Avoid steep slopes, as they are difficult to plant and cause a regular erosion problem, resulting in mud or dust on driveways and walkway areas. Maximum desirable slope is three horizontal units to one vertical unit. If space does not permit a gentler slope, use retaining walls (but do not forget the installation of a "weep" system to avoid hydraulic damage). Avoid terracing, as it creates a difficult situation from the standpoint of litter collection, grass cutting, snow removal, and erosion.

4. Water is as important to lawns as it is to humans; be sure that it is easily provided in adequate quantity. For other than small lawns, a built-in sprinkler system is typically the best grounds-care investment that can be made in new construction or renovation. The piping, preferably plastic, should be of at least ¾-in size, although risers may be ½-in pipe. Where possible, all lines should be drained to the main valves, and all lines sloped to a drain, but a minimum of drains should be

used. Again, where possible, sprinkler lines should be looped to eliminate dead ends which may become clogged. The valves and drain should be encased in a suitable valve box located about 2 in below sidewalk level to provide lawnmower clearance, except in the case of concrete boxes which should be next to the sidewalk and at its level. To avoid operating difficulties, valves should be of the globe type, and should have a union installed in the piping close to the valve to permit easy removal. Be sure that sprinklers will not dump water into a window well, and that water is not sprayed closer than 6 ft to a building. The control system should permit sprinkling various sections or areas and not necessarily the entire coverage at one time. An automatic timer should be used for turning sprinklers on and off, so that, for example, they operate for 15 min each evening.

5. Avoid the planting of small or irregularly shaped grass plots, as these are difficult to mow and often require a great deal of hand work. (See Fig. 8.) Injudicious arranging of small grass plots can add many man-hours to the maintenance work load. It is best to provide for continuous mowing and avoiding sharp turns around plantings, building corners, and other obstructions.

6. To eliminate hand trimming, use masonry or concrete mowing strips against buildings, under fencing, or at walls. Similarly, around trees use metal edging or grass barriers. This will also reduce the damage of trees from mowers. Further handwork should be eliminated through good design by keeping lawns clear of obstructions, such as plant beds and shrubbery located in "islands," which tend to disrupt continuous mowing patterns. Further handwork can be avoided through the use of flush paving around such obstructions such as sign posts, utility poles, fire hydrants, and sewer vents.

7. Signs, chains, and barriers have very little value in keeping people off the grass. Proper design of walkways as described later and the careful use of landscaping yield the best results.

All the problems typically associated with grassed areas—replanting, watering, mowing, fertilizing, litter removal, insect control—are eliminated or alleviated through the use of man-made turf substitutes. Several approaches may be taken in this regard,

36   Building Design for Maintainability

Fig. 8. Numerous small planting areas make grounds care difficult and expensive.

such as the green-colored ceramically coated granules of rock fixed in a polymer matrix, or artificial turfs. These are particularly good for relatively small areas that would be difficult to mow or to maintain without hazard (such as traffic islands, or roadway medians). Preparation should be made, however, to handle the maintenance problems associated with these substitutes.

Plantings are the crowning glory to landscaped grounds, and these can range from a simple row of flowers to a formal garden; from a couple of bushes at an entrance to a miniature forest. For plantings, use the following guidelines:

1. Use plantings which are indigenous to the area. Suitable information can usually be supplied by an experienced nurseryman or a county agent.

2. Give serious thought to the future costs involved in maintaining a planned flower garden. Many organizations have replaced flower gardens with evergreens and tree plantings in order to reduce their high maintenance costs. Inside buildings, consideration should be given to artificial flowers and plantings.

Actually, nothing can take the place of natural flowers, but their cost should be fully recognized in advance.

3. Carefully select the number of trees and shrubs to be planted; the tendency is to plant too many. A congestion of plantings means a lot of expensive gardening time; it also means a limitation on the size of mowing equipment and therefore more expensive lawn care.

4. Do not plant trees and shrubbery where roots will work their way into plumbing drain lines and where watering can soften the earth near the building foundation.

5. Probably more trees are damaged by being struck or scraped with lawnmowers than from any other source (Fig. 9). Nearly all this damage can be eliminated by providing a

Fig. 9. Where walkways are not provided, but are needed, they will be worn as paths. The trees will probably become damaged by mowing equipment since the grass grows right up to the trunk.

12-in unplanted space around trees where hand mowers will be used, and a 24-in or larger space where rider-type mowers will be used.

6. Where automobiles can be parked at curbs, or where trucks are used, set shrubs and trees back at least 6 ft from the curb to avoid damage to the plantings.

7. Where a building is designed to completely enclose a courtyard, provide methods to properly service the plantings, lawn, walks, and exterior building surfaces facing the courtyard. For grounds care and other heavy equipment, a passage should be designed through one of the ground floor areas providing doors of adequate width for the equipment, and using a durable floor material such as terrazzo or quarry tile on a firm foundation.

PAVING

The grounds are intersected and otherwise cut up with paved walkways, drives, parking areas, and the like; their treatment is important not only aesthetically, but from the standpoint of their own maintenance as well as that of the landscaped areas and even the building itself.

The basic choice for paving material is between concrete and bituminous asphalt, concrete being more lasting but having a higher initial cost. In either case, minimum maintenance requires application over a well-constructed base, and coating with sealers. Asphalt should never be used on a grade of more than 10 percent since it cannot be compacted properly on such a slope.

The size of grounds care equipment should be anticipated in the design of walkways, roads, ramps, and entrances, so that these will be sure to be wide enough.

Bear in mind these particular items:

1. Since most buildings are arranged in a rectangular fashion, it is normal for all walkways to be placed in a similar rectangular configuration. A year or so later, traffic patterns will have been worn in diagonals, and this unsightly condition may require additional walkways. Experience has indicated that

people will normally follow walkways if they are arranged with a consideration for foot traffic—and this would also typically include diagonal walkways. In campus and other heavily-trafficked situations, walkways should be a minimum of 8 or 10 ft wide. If usage is intense, the entire area may require paving, perhaps interspersed with plantings or planters. Since pedestrians tend to "cut corners" at the intersection of walkways, thus destroying grassed areas and plantings, walks should contain a 5-ft minimum radius at intersections.

2. Provide for a crown or a 2 percent slope (the latter is easier to construct) in all walkways to prevent water from collecting in the center. The accumulation of water can cause slipping hazards (especially on freezing) and deterioration of the surface (again, especially on freezing), as well as the tracking of mud and water into buildings.

3. Avoid the use of brick, stone, or wood feature strips in concrete walks or drives as they tend to provide innumerable openings in which water can collect and, on freezing, break up the paving. Also, they provide crevices in which weeds and grass can grow, which presents another type of maintenance problem. They can become obstacles to the use of snowplows, or brooms, and can present safety hazards when the features become raised through ground movement or freezing.

4. In heavy snow areas especially, avoid dead ends in walks and drives, as such areas cannot be suitably cleared of snow by powered equipment; also, power sweeping is difficult.

5. Barricade any paved or surfaced area, such as decking or walkways, which is not strong enough to support vehicles. This barricade may consist of plantings, curbs, fencing, or other means.

6. Near driveways, eliminate any construction which can be damaged by vehicles; columns should be set back from the curb, and walls should be protected by corner guards.

7. If a specified area is set up for parking semitrailers which are not attached to their tractors, and the parking area is paved with asphalt, place a special concrete pad or strip in the paving so that the steel wheels supporting the trailer are resting on the concrete rather than on the soft, yielding asphalt.

8. Equip loading and parking areas with a storm drain to avoid ponding, but also grade these areas so that if the storm drain becomes stopped up the water will flow out of the area before entering the surrounding buildings.

9. In cold-weather locales, consider the use of built-in snow-melting equipment to minimize manual labor, decrease the trackage of soil into the building, and eliminate accidents. The principal varieties include coils imbedded in the concrete walkway, the heat being provided by steam or hot water; electric resistance wires, also built in, although these may provide more technical problems; and electric radiation from an overhead canopy, which may be used independently or in conjunction with one of the other systems.

Fig. 10. A ramp, rather than the steps, would have made waste removal and other activities much simpler and safer. (*Compackager Corporation.*)

10. For curbing, consider granite, concrete, and asphalt, in that order. For real durability, only granite resists both the ravages of weather and snowplows.

11. Provide ramps at curbs for wheelchairs, moving of office equipment, and of powered maintenance equipment. (See Fig. 10.) Install one wheelchair ramp for each building used by the public, or provide at least one ramp for the use of dollies and package carts where luggage or other items must be rolled about. A wheelchair ramp should be at least 4 ft wide and should not exceed a 7.5 percent gradient or be more than 30 ft long. The ramp should be continuous throughout its length; a platform or level area should be provided outside the entrance door to a building.

EXTERNAL FEATURES

A variety of features, at first glance perhaps seemingly unrelated but in actuality part of a comprehensive program of maintainability, should receive study:

1. Just as there is a need for custodial closets on the inside of buildings, there is a need for a grounds-keeping closet on the outside. (See Fig. 11.) Each major building should have its own closet at ground level, opening to the grounds; an opening to the interior of the building is not necessary. (A convenient location is often next to an electrical or mechanical room.) Where buildings are close together, one closet might serve more than one building. The size would depend upon the nature of the grounds-care activities, but typically from 6×8 to 6×10 ft would be suitable and could be used for the storage of such things as raincoats, hand mowers, trimmers, hose reels, chemicals, tools—and lunch boxes!

2. Immediately next to buildings, it may be desirable to provide a pea-gravel splash area to prevent mud from spattering on the lower few feet of the building. The pea gravel should rest on a plastic film to prevent the growth of weeds, and the gravel should be separated from the lawn areas.

3. For those organizations that use bulk containers for waste collection, the location of the containers is a problem. (See Fig.

Fig. 11. This unsightly exterior area could have been walled in as a storage room.

12.) Ideally, they should be placed at a loading dock under an overhang, which also mades it easy to put the waste into the containers. On the other hand, where the container is next to a building, there is often a masonry enclosure, but such an enclosure is easily damaged and is time-consuming to repair, so plenty of room should be left around the sides for movement of the collector. Also consider the use of a fence-type enclosure that is easily repairable and that might give rather than break on impact. Avoid shrubbery enclosures, as they frequently become entangled with the litter. In some cases it is possible to put containers inside a room in the building, which permits the use of chutes directly from each floor to the container, thus saving many man-hours in trash removal. (Of course, the chutes should be equipped with sprinkler heads and a fusible-link-actuated door on the bottom for fire protection.)

4. Consider lighting parking lots, walkways, and other exterior areas at a minimum of 2 fc. This will be of considerable help in security control and the minimizing of vandalism.

Fig. 12. Proper exterior space for waste control systems and their necessary utilities are part of initial planning. (*Ross Engineering Company.*)

5. Especially where only a few steps are involved, consider the use of a ramp to eliminate the problems of step maintenance and cleaning, as well as to eliminate a safety hazard. The ramp further provides wheelchair, cart, and equipment access, as previously mentioned.

6. Provide adequate facilities for loading and unloading trucks, to prevent damage to adjacent areas. Such facilities should be covered, and ice-melting devices should be considered. If the drive slopes downward toward the truck dock, then a suitable drain should be installed. The loading dock, again, is an excellent area for a bulk waste-disposal container or compaction device, in a permanently marked position. Provide a rubber

or other similar type of dock bumper so that when trucks approach the dock with excessive force the shock will not be transmitted to the building structure to cause cracking of plaster, breaking of windows, or damage to the dock itself. A dock-leveling device can avoid damage to the dock, facilitate handling of materials, and avoid accidents.

7. The removal of weeds and litter along a fence line is a time-consuming job that is required quite regularly. The weeding problem can be eliminated completely and the littering problem can be greatly simplified by installing a concrete strip—about 2 ft wide and 4 in thick—under the fence. The strip further permits the use of a power mower (assuming the concrete pad is below the grass level) without any hand trimming needed, and it permits the use of a vacuum machine for the removal of litter.

8. So much time is spent repairing and replacing damaged outdoor drinking fountains, as in parks, school yards, public-building areas, and the like, that they should probably be eliminated altogether. If this is done, the drinking fountains within the buildings should probably be located as near the outside doors as possible.

SOIL ENTRAPMENT

Of all the important considerations in designing buildings for better maintainability, soil entrapment would have to be at or near the very top of the list. How much easier it is to prevent soil from entering and being spread around the building than it is to clean it up from many thousands of square feet inside! The subject offers a number of possibilities, some of which are covered in other chapters and in other sections of this chapter. To really get the job done, a single system of trapping soil at entrances is generally not sufficient; rather, at least two complementary systems and sometimes three are required for severe situations. For example, in bad-weather areas a visitor entering a building may first use a walkway containing a snow-melting device, then cross a grating which removes much of the gross soil from the shoe soles, then finally walk across a matting which removes most of the remainder of the soil. Let us look at some of our possibilities.

1. If walkways are covered at least 25 ft from the primary entrance, interior floors will be much better protected from the ravages of soil and water, as the length of covered walk will act as a soil entrapment device rather than as a soil-producing surface.

2. Where a walkway, apron, or lanai is made of a smooth surface, soil tends to be carried into the building. On the other hand, if the surface is roughened, such as through the use of certain natural rough stone or of pea gravel imbedded in concrete, it will then act as a soil trap (of course, periodic soil removal is required so that the soil trap does not itself become a source of soil to be tracked into the building).

3. For many buildings, the best candidate for investment may well be a grating with bars spaced about $\frac{1}{10}$ in apart (to prevent the catching of narrow heels and the like), with a catch pan beneath (Fig. 13). The grated area should be large enough to assure that each foot will come into contact with it at least once during the entry; this would dictate that the grating be at least 3 ft in depth and preferably 4 ft or more. Where the bars are constructed of steel, rusting should be expected; aluminum or stainless are much better. Some varieties have a replaceable plastic tip that is itself crowned and runs the length of the bars, which gives good service in acting as a squeegee in removing soil. Another variety has a set of vibrating brushes between the bars which mechanically clean the soles of the shoes. The vibrating is actuated by a contact switch when weight is being applied to the grating surfaces. Of course, the brushes require adjustment and replacing and this device should only be considered where the most severe conditions pertain or where an unusually high quality of floor maintenance is required—and perhaps where there is no room or condition that would permit walk-off matting beyond the grating. Where a good deal of water is collected in the catch pan, some form of discharge should be provided, preferably a weep system; but catch pans connected to plumbing drains tend to clog. An exterior stair landing is

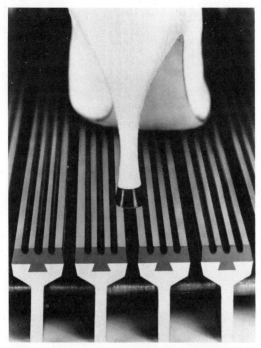

Fig. 13. Most efficient soil entrapment is provided by a grating with a catch pan beneath; but the traffic should be at right angles to the bars. (*Pedigrid Company.*)

a good place for a grating, especially if there is a canopy overhead, and this condition can also be provided in renovating buildings by making an extension to the older stair landing. Even better, grating should be placed between double sets of doors, with walk-off matting or replaceable carpet area beyond the second door. Circular gratings are even available for revolving doors.

    4. Carpet itself acts as a good soil trap, not only at entrances but in such locations as stair landings and elevators. This subject is covered in detail in Chap. 4.

    5. Matting is available in numerous sizes and various materials. The least desirable is the coco-mat, since it is very difficult

to clean, tends to unravel, and is unsightly; further, soil tends to penetrate through the mat and damage the flooring beneath. When replacing carpet, some people save the less-worn portions for use as matting, after binding or taping the edges to prevent raveling. (In a typical room, some carpet areas will be worn through while others will show almost no wear at all.) Matting is now also being made of synthetic chemical fibers and extruded chemicals, as well as impregnated cotton fiber. Nylon has the advantage of being easily cleaned, and impregnated cotton may be cleaned in a commercial-type laundry, but many people use a contract rental system where the soiled mats are picked up and cleaned and treated ones are left in their place. Typically, the larger the matting the more successful it is and also the less likely to be stolen or to cause slipping or tripping accidents. (Be sure that the corners lie flat!) Preferably, matting should be recessed so that the wearing surface is level with the floor. Special forms of matting include the tacky mat and sponge mat. Tacky mats are comprised of a number of layers of film with a sticky surface, somewhat like pressure-sensitive tape. They are used, for example, at entrances to superclean rooms where dust count is an important factor. Sponge mats consist of a sheet of spongy material in a stainless steel pan, the sponge being impregnated with a disinfectant solution. These are sometimes used at entrances to hospital operating suites or other areas where bacteria control is important.

6. A runner is simply a long mat. Runners can be purchased in various widths and lengths; typically a 4-ft width seems to do most jobs. Again, a recessed area should be provided in the design stage. The recess also tends to act as a catch pan for the soil that moves to the side of the runner. One type of runner is motorized to provide new surfaces for catching soil, and self-cleaning runners will undoubtedly become a regular feature of buildings in the future.

7. For special situations, such as industrial superclean rooms (white rooms) an air bath can remove much loose soil from the body and clothing, as a laminar flow of air moves down from an overhead source and is exhausted through a vacuum below a foot grating. (See Fig. 14.)

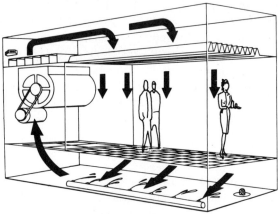

Fig. 14. An air curtain allows unobstructed passage and eliminates most door maintenance, as only security doors are required. An integral part of the unit is a floor grating. (*The Stanley Works.*)

EXTERIOR BUILDING SURFACES

The focus of attention on any property, naturally, is the structure itself. First impressions being what they are, not only the interior of the building, but the very reputation of the management will be judged by the appearance of exterior surfaces. A positive first impression should be maintained, without excessive cost, for as many years as the building will be occupied. A discussion of exterior surfaces follows:

1. Ideal exterior surfaces are polished stone (such as granite), stainless steel, and glass—they are easily cleaned, durable, and resist graffiti. Use these materials up to a minimum height of 7 ft. Use anodized aluminum for exterior trim to avoid tarnishing and deteriorating.

2. To avoid surfaces that require regular cleaning or painting, specify exterior walls of brick, cast stone, natural stone, or any other permanent type of material. Wood should be judiciously used, or used sparingly. Figures 15 to 18 illustrate some of the more easily maintained surface materials.

3. Avoid porous stone, such as limestone or some varieties of marble, both in facing buildings and for other purposes. Stains and watermarks in the stones are difficult if not impossible to remove. For example, the metal clips that hold the stones to the buildings, even though galvanized, may eventually rust, and these rust stains may completely penetrate the stone, permanently defacing the structure. Porous stones and concrete can be somewhat protected through careful application of solvent-type sealers; these should be of the penetrating type, which give the best protection under these conditions, and surface sealers may tend to discolor in sunlight.

4. Precast concrete, molded plastic, or stamped metal panels should be formed so that no horizontal surfaces are exposed, as these tend to collect litter and soil which is normally inaccessible for cleaning. Such surfaces should be beveled—typically at 45°—to avoid this problem.

5. In tall buildings, the exterior masonry wall sometimes fails due to compression caused by building shrinkage. To overcome this problem, lintels supporting the brick, typically at each

Fig. 15. The mottled green marble appears almost unmarked after many years of heavy service. A white stone under the same conditions would have become most unsightly.

floor level, should be provided with a compression layer of material that is a minimum of ⅜ in thick. Since steel is the most commonly used material for lintels, take extreme care to enclose the lintels within the masonry to prevent rust stains on the building exterior. In general, any steel not thoroughly imbedded in masonry—where any part of it is exposed to the weather—becomes a problem, as it is completely inaccessible for maintenance painting. (We refer here to standard steels rather than to the specialized materials which depend on controlled corrosion for their protection and appearance.)

6. Avoid designing the exterior of the building as a "natural ladder"—some forms of decorative masonry create this effect—since this can lead to vandalism that might otherwise not occur. Figure 19 shows an example of good, exterior design.

7. Outside stairways should be made of concrete or stone; avoid steel, as it requires painting. The treads should provide an abrasive system, such as inset carborundum strips, to prevent

Fig. 16. Vinyl siding is growing in popularity. It is nonflammable, nonrusting, and easily cleaned. (*Certain-teed Products Corp.*)

slipping as well as damage to the tread. Where freezing is not a problem, cast-iron nosings with abrasive impregnation is successful (freezing might tend to separate such a nosing from the concrete). Treads should be sloped to drain off the water. Handrails should be installed at the time of construction for greater strength and to avoid damaging the treads through later installation.

8. Scrupulously avoid the use of wall-mounted signs comprised of individual letters mounted directly on the building exterior. Attempts to clean the letters will result in the sign and the building becoming damaged. Further, when letters fall off or are otherwise removed, their replacement becomes tedious and costly—and at times impossible. Avoid any form of sign

52   Building Design for Maintainability

Fig. 17. Weathering steel eliminates all painting. (*Bethlehem Steel Company.*)

Fig. 18. Porcelain enamel siding eliminates painting and minimizes exterior cleaning. (*Bethlehem Steel Company.*)

Fig. 19. The simple exterior lines of this technical school lead to long life and low maintenance costs. (*Ellerbe Architects, St. Paul.*)

that requires polishing, such as a brass embossed sign, especially if it is mounted on a door or wall. An ideal sign is one made of plastic, with the lettering on the inside or backside (not exposed to the weather), and possibly lighted from the interior. Such signs are easy to clean with neutral detergents.

Even though vandalism often seems to occur in cycles, it will always be a problem and should always be considered in exterior building design. A building with an irregular configuration provides concealment for vandals and is thus apt to require much more money for repairs than a building with simpler lines. This consideration would be most important for buildings that are not in continuous use, such as public schools, certain industries, and shopping centers. Where the irregular configuration is desired, or even mandatory because of the function of the building, the use of protective fencing and additional lighting becomes a likely requirement.

ROOF

Moving now to the top of the building, we are involved in a discussion of a surface that protects all that lies beneath. Improper provision for roof maintenance can cause an annoyance here or there, or can lead to a wholesale disaster in terms of both property damage and lost time; proper design, on the other hand, contributes to the maintenance of both the roof itself and the rest of the building, as shown in Figs. 20 and 21.

The built-up roof of any building presents a large area where maintainance is often required but seldom done. It seems that since the roof is generally out of sight it is indeed also out of mind. Many building owners have the attitude that their 10-, 15-, or 20-year roof bond is all they need for protection, but there has never been a roof bond that prevented leaks. The bond is merely a form of insurance which partially pays for repairs. Further, it guarantees only those repairs required because of ordinary wear and tear. It does not cover insulation, vapor barrier, and metal flashing, all of which are essential parts of the "roof system."

Fig. 20. This aluminum-overlaid plywood mansard roof combines ease of installation with low maintenance. (*Weyerhaeuser Company.*)

Fig. 21. This maintenance system avoids foot traffic and the damage it can cause to the roof. (*Mannesmann Engineered Products.*)

Some years ago, roofing materials manufacturers introduced the roof inspection and service contract. Of course, the existence of such a contract, just as the roof bond, does not prevent the roof from leaking. There is simply no substitute for good design, materials, workmanship, and maintenance. The contract does, however, provide for inspection of the roof by the materials manufacturer's experts in at least the second and tenth year of roof life to keep the building owners aware of their roof and the need for good housekeeping practices and preventive maintenance. Some contracts provide for inspection every year for the duration of the contract.

Another method of assigning financial responsibility for the roof system is the roofer's guarantee. Such guarantees are generally written for one or two years and require the roofer to repair leaks resulting from faults or defects in material and workmanship.

Careful consideration of the method used for assigning financial responsibility for the roofing membrane is important, but more important, to assure the integrity of the entire roof system, is the careful consideration of all its components and the factors affecting its performance. The designer must consider the structural deck, vapor barrier, thermal insulation, roofing membrane, aggregate surfacing (if any), flashings, expansion joints, vapor control and drainage, in order to assure an integrated system which will perform its intended function. Literature is available to the designer and there is also much available for building owners and maintenance personnel to assist them in preventing that much despised and costly roof leak. The scope of this book does not permit a comprehensive discussion of this complex subject, but the following items should be helpful.

Since protection from the elements is the basic purpose of the roof, we should immediately eliminate from our design a flat roof, where the problem in maintaining watertight integrity is well demonstrated by the difficulty—impossibility, in many cases—of obtaining guarantees. Water leaking into a bedroom closet in my parents' home, after much searching, was found to come from an almost invisible tear in the flat roof over 25 ft away from the point of damage! The many experiences of that type indicate the need for a minimum slope of $2\frac{1}{2}$ percent, with the slope directed to a roof drain.

In many cases, the biggest problem connected with the maintenance of built-up roofs is finding leaks so that they can be repaired. The use of aggregate surfaces may make this difficult. There are some cases when eliminating the surface aggregate and leaving the roof surface smooth could reduce the search and repair time.

For industrial installations, and some others as well, consider roofing materials which may have an almost indefinite lifespan, such as aluminum (providing the atmosphere does not contain chemicals injurious to this metal, and provided it is properly installed) or corrugated fiber glass. The combination of these two materials, with the fiber glass providing for natural light, is often used successfully.

Many exterior walls have been marred, and the ground eroded below, by the overflow of water at roof edges because of an inadequate number of roof drains. Calculations for drains should assume that a certain amount of clogging will take place, and that construction irregularities may actually put some of them out of regular use.

Guttering should be of adequate size to avoid clogging from foreign matter; this is true for the down spouts as well. Be sure to provide splash blocks at the foot of the down spout to prevent spattering the building wall—or, preferably, connect the down spout directly with a storm drainage system. For many installations, a continuous aluminum gutter (with no transverse joints) is best, as it avoids the usual problem of leaking joints, as well as the need for interior painting. For long runs of aluminum gutter, however, remember to allow for the greater expansion and contraction due to heat changes than takes place with galvanized steel. Where vapors are present that are corrosive to aluminum, plastic or stainless steel might be considered.

Some other items affecting the maintainability of roofs follow:

1. Where a step-back construction has been used, install a permanent ladder for easy access to each roof level—assuming that access is not possible through an elevator. These ladders, which should be of galvanized steel or aluminum, will facilitate the jobs of making periodic inspections and minor repairs as well as removing litter from roof surfaces. Although easy access from one level of a roof to another is desirable for maintenance purposes, easy access from the ground to the roof for properties not protected against vandals considerably simplifies building entry and escape. Retaining walls and trees close to a building can also contribute to this problem.

2. If the roofing is of the type that is easily damaged by foot traffic, construct walkways (or purchase tread boards from roofing manufacturers) for roof inspection and maintenance. This would be especially true around elevator penthouses and stairway entrances.

3. If you can, provide for a roof overhang 4 to 6 ft wide, as it offers considerable protection to the sides of a building. It leads to less chemical and other cleaning of the building surface, less window washing, and less damage to doors. Under

Fig. 22. Both the downspout and the wall require protection from damage by automobiles.

certain conditions, it can also reduce the amount of soil brought into a building.

4. Remember that where canopies are installed without drains, the dripping water damages the ground below to the extent that a trench is formed; the water in the trench, in turn, can find its way into the building. The drain should therefore be considered part of the canopy design and installation.

5. Connect the downspout to a steel sleeve or boot at least 5 ft above the ground (Fig. 22). The boot itself may need protection if it is in a position to be struck by trucks, although such a condition would indicate the desirability of relocating the downspouts. In cold-weather climates, where maintenance problems due to water freezing in gutters occur repeatedly, consider downspouts on the interior of the building. Such downspouts would, of course, have to be located where they would not be damaged. The spout opening on the roof must be protected from clogging by a cap strainer. In cold-weather areas exterior downspouts should be provided with leaf cables.

6. Use flashing and related roofing materials of copper,

Fig. 23. Aluminum soffit and matching fascia will not rot, split, or rust. (*Kaiser Aluminum Company.*)

stainless steel, or aluminum. Ferrous materials create excessive maintenance costs. Do not imbed aluminum in concrete or masonry, or permit it to make contact with ferrous metals other than stainless steel, in order to avoid its deterioration. Fasteners used in aluminum must be of either aluminum or stainless steel. (See Fig. 23.)

7. Wherever a wire or bar penetrates the roofing material, provide a pitch pocket to prevent leakage (a pitch pocket consists of a metal sleeve, flashed to the roof covering, and filled with roofing pitch), but avoid pitch pockets for columns and other structural members penetrating the roof—for these, a pedestal should be constructed and flashed with a cap. To prevent situations that lead to leakage, do not permit objects spaced closer together than 18 in to project through the roof unless each is provided with its own flashing; and nothing should pierce the roof closer than 24 in from a cant stripe or fire wall.

8. Where ventilator pipes or other projections come through the roof, attach a deflector hood or skirt near the roof level so that water will be shed away from the curb mounting (a

properly sloped roof and good flashing design might well eliminate the hood or skirt).

9. Provide lead caps on horizontal top joints of stone copings or overhanging stone corners. Do not permit flat surfaces on tops of walls and parapets or sills, which, if they do not drain readily, will hold water and allow it to penetrate to the interior of the building. Of course, if possible, avoid parapet walls altogether, or any other decorative roof feature.

10. Be sure that mechanical equipment on the roof is protected so that it will not freeze in bad weather. It would be preferable, of course, to avoid roof-mounted equipment entirely by having it in a mechanical room under the roof.

A special consideration is roof insulation, although the benefits derived may be more in terms of operating costs than in maintenance or repair, but still the latter can be seriously affected. A well-insulated roof reduces the heat loss to such an extent that a different form of heating may be considered (especially important in large-roof-area buildings such as shopping centers). Thus, for example, it may be possible to change from an oil-fired heating system to electric-resistance heating, perhaps thereby considerably reducing maintenance costs (as well as operating costs, in some locales). In general, use inorganic materials for both insulation and roofing felts, as organic materials disintegrate if they are saturated with water.

BIRD CONTROL

Bird droppings have turned many a beautiful building into an unsightly mess. There are numerous ramifications to this problem, involving such facts as the carrying of disease by some birds, the pleasure that some people derive from feeding birds, the actions of conservation organizations, the designation of bird sanctuaries, and so forth. Bird control is variously approached through killing, repelling, trapping, and—the subject of this book—building design.

Although it is difficult to imagine a building that could be designed with no areas on which a bird could light, it is certainly possible to design a building that minimizes such areas. This is the prime factor in eliminating the bird nuisance. The trend in building design has been toward simplification, and elimination of cornices, friezes, statuary, and the like, and this of course has tended toward simplification not only of the bird problem but of other problems as well.

Where buildings already have numerous surfaces on which birds may alight, or where the design requires such surfaces, such as in churches, wire netting—of nonrusting materials, such as aluminum, copper, or stainless steel—with openings of less than 1 in, may be used. On narrow ledges, sheet metal can be installed at a 45° angle so that the surface is one upon which birds cannot alight. There are manufacturers of spiny or needle-tipped contrivances which are meant to be installed in relatively small areas, such as over doorways or windows. Sometimes birds cope with this problem by collecting refuse to fill in the spaces between the wires or spines, so they may need occasional refuse-removal attention. (Of course, these devices can attract and hold refuse that is simply being blown about, as well.)

Electrified wires, such as the type used in prison compounds and to keep strangers off fenced property, have been attempted, but the results have been unsatisfactory, principally because it is too difficult to maintain the insulation of the wire from the building—one wet stick, for example, will ruin it. Where it is possible to have access to the wires to keep them uncluttered, then this can be a successful system. The electric charge should not be one that kills the birds, but that merely keeps them away.

CHAPTER FOUR

Floors, Elevators, and Stairs

From a surface maintenance point of view, the most critical item in any building is the floors. For example, in custodial maintenance typically 40 to 60 percent of the cost is involved in floor care. Floors receive as much as 90 percent more wear than any other building surface—and the floor is usually the most noticeable surface in the building. Design for floor maintenance is a most important aspect of building planning, with consideration being given to soil retention, soil removal, stain removal, life, ease under foot, and appearance. Elevators and stairs are made a part of this chapter, since in a sense they are special types of floors: the elevator can be considered a movable floor, while the stairway can be thought of as an interrupted ramp.

At times it becomes desirable to replace floor tile in certain areas where there is excessive wear, such as around drinking fountains, at the bottom of stairwells, at the entrances to elevators, and the like. Attempts to match the original tile color are usually unsuccessful—even if some of the original tiles have been kept on hand as spares—because of discoloration due to light. Even the thickness is different because of the worn tile adjacent to the removed portions. In these cases, it is best to deliberately choose a contrasting color. For example, if the general floor is tan, then the replacement tile might be maroon or green. The replacement area, of course, would consist of a complete rectangle. (The same principle is true for terrazzo.) It is a good practice on initial installation to order a small percentage of extra tile in a solid color based on the field, or major color, of the mottled or speckled tile being used; then the solid-color tile can later be used for a rectangular replacement area and will blend well in color.

Where explosive vapors may occur, such as in hospital operating rooms, certain chemical laboratories, and some hazardous production areas, conductive floors are provided so that a static electric charge will not build up on insulated surfaces which would lead to a spark that might detonate the vapor. Various surfaces are used for this purpose, such as conductive vinyl sheeting, conductive ceramic, and conductive terrazzo; of these, the last is the most durable and successful, since no finish should be used to protect the surface as it may provide an insulation. An automatic electrical-resistance-testing system should be installed to be sure that the floor is properly conductive.

High-pressure laminated plastic materials, specifically manufactured for the purpose—similar to General Electric's Formica, and Westinghouse's Micarta—are ideally suited for pedestal floors, which are now used exlusively in data processing areas. Such surfaces, although rather expensive, avoid the necessity for all finishes and require only damp cleaning for maintenance. Their wear qualities are superb.

Normally spaced floor drains should be provided for areas which are usually dry, but that are subject to periodic wetting or hosing. For floors which are continually wet, provide trench drains with grating covers (make sure these covers are strong enough to carry the heaviest vehicle which might be used in the area). Of course, do not forget the floor sloping. The bottom of the drainage trench itself should be sloped $\frac{1}{4}$ in/ft to carry off the water.

Again, where floors are generally wet, consider concrete curbs which are poured integrally with the floor slab at all wall junctions, in order to prevent leakage from upper floors or service levels down through the construction cracks that would otherwise be between the side walls and floors. Also provide a raised curb or sill at doorways or entrances to electrical control rooms, with ramp approaches on both sides for easy movement of equipment.

A wide-sweep curve from floor to wall provides initially an attractive appearance (and leaves no corners to collect dirt and litter), but if the curve ends more than 6 in above the floor there will be no practical means of maintaining this surface. An airport concourse, for example, was seen to have a white vinyl-asbestos floor curving up the wall to a height of about a foot. The curved area was badly marked by luggage and carts, but there is no floor care equipment that can scrub more than 6 in high; if this area is ever to be cleaned, it will require the development of special equipment and special operator skills.

If aisle markers will be a permanent feature of the floor, consider emplacing them on a permanent basis. Aisle markers such as paint stripes, plastic disks, and adhesive spots need constant replacement and maintenance, yet aisle markers are a definite advantage in terms of traffic flow, location of equipment, storage, and definition of areas of responsibility. Permanent marking might consist of a contrasting color of concrete or terrazzo, embedded ceramic blocks, or metal strips.

Avoid the open-slot type of expansion joint in floors as these tend to fill up with soil and waste and are also a safety hazard. The expansion joint should be a type which has a sliding member, or at least a filler material for the slot so that the surface is essentially level.

Avoid pouring a concrete slab in direct contact with a masonry wall, as movement of the concrete will tend to crack the wall. Specify the use of expansion-joint material made for this purpose.

CARPETING

Unquestionably carpeting has a great many benefits to offer, and the great swing toward carpeted floors would not have taken place were this not so. On the other hand, it is not a panacea and it is certainly not the best flooring material in all cases. Typically, carpeting is a good investment where it is selected for the proper type of area, and where the correct type of carpet is specified. It is an excellent choice for lobbies where soil entrapment devices are in use so that mud and dirt are not tracked directly onto the main carpeted area; for elevator floors and stair landings as soil entrapment devices themselves; for dining rooms and cafeterias where it best exhibits its quality of sound control; in public, office, dormitory, school, and, as a matter of fact, most general types of areas where there is not very heavy traffic and where a great deal of soil is not tracked in. In patient and treatment areas of hospitals and other medical care areas, carpet has been used successfully, but there still remains a question concerning cross-infection control, and the control of odors and spillages.

When it comes to carpet selection, special care must be given to the construction, coloration, and other qualities of the carpet itself. Absolutely to be avoided are cotton carpeting, cut pile, deep pile (over 3/8 in); light colors such as white, yellow, or gold; and solid colors which immediately show all traffic patterns and any stain. The possible use of carpeting must be analyzed on the basis of economic and other factors (such as sound control, appearance, ease under foot, public relations) for each potential use.

Some special considerations:

1. Where wheelchairs or cart traffic is anticipated, carpeting must be of dense, low-pile structure.

2. For hotel rooms and other lighter-use areas, consider three-color shortcropped synthetic shag. It permits easy replacement of worn areas and is relatively easy to clean.

3. In office areas, provide large casters of the ball type for movable chairs, carts, and equipment, and use dense carpeting. Desk pads are not too successful, and are unsightly.

4. Do not permit excessive mixing of carpet and other types of floor surfaces, which greatly compounds the cleaning problem. It would be better to extend the carpet to include all adjacent areas (Fig. 24).

5. For certain areas, consider carefully the flame characteristics of carpet.

Ideally, carpet should be obtained in a four-colored tweed, tight-loop pile of continuous synthetic filament, or a nylon-wool

66   Building Design for Maintainability

Fig. 24. Mixed carpet types provide interesting treatment while avoiding the problems of mixing carpet with finished floors. (*Reynolds, Smith & Hills, Architects, Jacksonville, Florida.*)

blend, with a pile height of about $\frac{1}{4}$ in on an impermeable backing membrane. A typical specification for a practical material follows:

	Wool	Polyester	Nylon or olefin
Gauge (tufted)	$\frac{1}{8}$ in	$\frac{1}{8}$ in	$\frac{1}{8}$ in
Pitch (woven)	216	216	216
Rows/inch	7–10	7–10	7–10
Pile height (inch)	0.250	0.250	0.250
Ply	3	3	3
Pile weight/square yard	45	40	30

For tufted construction:
    Primary backing—3.5-oz polypropylene
    Secondary backing—7-oz jute
    Tufted bind—15-lb minimum

FLOORING TYPES

In addition to carpet, there are numerous other types of flooring available now, and others coming on the market regularly. There are so many factors that relate to proper floor choice for maintainability that no generalization can be made. A review of the more regularly used floor types (alphabetically arranged) should be helpful.

Asphalt Tile

One of the least expensive of all floors is asphalt tile, but it is also one of the poorest performers. The surface indents very easily and does not recover, such as from the pressure of chair legs; it is also quite susceptible to damage by solvents, including some of the materials used for dustmop treatment. The resistance to wear is low. If resilient tile is to be used, a much better investment is vinyl asbestos.

Ceramic Tile

Ceramic tile is one of the most durable and trouble-free floor surfaces available (Fig. 25). It is ideally suited for areas where a good deal of moisture is present, such as in food service or food processing areas, some chemical manufacturing, restrooms and locker rooms, shower and change areas, laundry rooms, custodial closets, and swimming pool aprons. The cement grout between the tiles is much more porous than the tiles are, and therefore should be protected with sealers, just as would be used for terrazzo. On the application of such sealers, all surface residue should first be removed from the tile, as otherwise the sealer may not adhere properly to that surface. Ideally, the grout should be of a colored variety, such as caramel or gray, so that soil or the formation of mold are not so apparent. In wet areas, obtain ceramic tile with carborundum imbedded to avoid slipping accidents.

Concrete

Concrete is probably the most common of all commercial or industrial flooring material because of its initial low cost and

68   Building Design for Maintainability

Fig. 25. Quarry tile can combine attractiveness, ease of maintenance, and durability under heavy traffic conditions. (*American Olean Tile Company.*)

durability. Its surface density and durability can be increased through the use of membrane sealers during its installation, of solvent or water-emulsion sealers in normal use, or of densifying materials (including various chemicals, minerals, and even steel particles), that add to its cost, during installation. Some of the newer epoxy-type patching materials permit even thin sections of concrete to be repaired successfully. One form of concrete has a coloration mixed in, although this is never fully successful because of traffic patterns being formed as the surface of the concrete roughens. Painting the concrete is also not usually a good idea where there is any traffic at all because of the great amount of time required to keep up its unmarred appearance. (This can be successful in boiler rooms and other machine areas where there is very little traffic but a high appearance is desired.) Ideal applications of concrete floors are for industrial areas, warehouses, mechanical areas, storage, utility areas, and locker rooms and restrooms where high-quality appearance is not required. One of the advantages of concrete is that, to upgrade its appearance, it may be covered over with such surfaces as ceramic tile, resilient tile, or carpet. The use of abrasive particles can provide a nonslip surface, such as in areas that are frequently or constantly wetted.

Where a floor is to be hosed down regularly, such as in a meat-packing or food-processing plant, the floor should be sloped a minimum of $\frac{1}{4}$ in/ft for drainage. This sloping can be done in areas or sections, making the working areas and passageways the high points. Also, slope floors away from stairways, doorways, elevators, and other openings.

Where floors are apt to become wet, avoid a steel-troweled finish on concrete floors; a float finish will be much less slippery. Concrete may also contain nonslip additives, such as carborundum particles, which are preferable to any sort of nonslip coating or covering. (Ceramic tile and vinyl-asbestos tile may also be obtained with carborundum particles embedded for antislip purposes.)

Use a waterproofing membrane under concrete floors that are poured on grade; this ensures slower and better curing of the concrete and keeps ground moisture from rising through the slab, thus slowing surface disintegration and equipment corrosion.

Cork

Cork, being a special form of wood, provides a special type of wood floor. Its principal advantage is its sound-absorbency; and to some, it is a very attractive surface. It is, however, a difficult floor to maintain, and many errors occur in its maintenance, such as the application of water. In practice, carpet provides all the advantages of cork (other than its unique appearance) at much more economical maintenance costs. Never apply cork tile to a concrete floor on or below grade.

70 Building Design for Maintainability

Fig. 26. Sheet vinyl flooring is especially desirable for hospitals, as it eliminates cracks and crevices that might harbor germ-laden soil. (*Goodyear Tire & Rubber Company.*)

Homogeneous Vinyl

Pure vinyl flooring, either in sheet or tile form, is rather expensive and is not generally a good investment except for problem areas. For sheet vinyl, this might be where there is spillage (Fig. 26); for tile floors, where there is very heavy traffic and considerable soiling. Contrary to manufacturers' recommendations, pure vinyl should be given a floor finish for protection and best appearance, except in the most lightly trafficked areas. Minimum thickness should be $\frac{3}{32}$ in.

Impregnated Wood

A new generation of wood floors involves the impregnation of the fibers with plastic materials such as acrylic. These acrylic/wood floors, for example, are being successfully used in shopping center malls, building lobbies, and other heavily trafficked public areas. The further development of these materials will see their greater use in the future, as very little maintenance is required.

Linoleum

The principal advantage of linoleum flooring is its low initial cost. It also eliminates the joints found with resilient tiles, which is an advantage where moisture is present. However, its appearance leaves a great deal to be desired and it is not normally used except in low-class situations.

Marble

In theory marble should be a good flooring material, just as terrazzo is. In practice, however, the formation of traffic patterns and other problems provide an uneven appearance that is difficult to control. Further, cracking due to building settlement or damage to the floor is unsightly. If the appearance of marble is desired, consider it for wall material.

Mosaic Tile

Decorative mosaic tile, a form of ceramic tile, can provide an extremely attractive yet very durable surface for heavily trafficked areas, such as vestibules, entries, lobbies, and ground floor corridors. (See Fig. 27.) For best service and lowest maintenance, the portland cement grouting should be properly sealed, or a grouting of synthetic plastic material may be used. In either case, the grouting should be colored to blend in with the color of the mosaic tile.

Plastic Laminates

A variety of floor covering material is available that consists of numerous films of plastic (such as polyurethane) applied on a concrete subfloor. Some intermediate layers can be speckled or colored to give an attractive appearance, and the final top surface left in the desired condition of smoothness or roughness. Although the theory is excellent (providing a smooth, continuous surface without cracks or joints, resistant to stains and chemicals and easily cleanable), in practice the results have been mixed. Much of this has to do with the skill of the applicator, but

72 Building Design for Maintainability

Fig. 27. Decorative or mosaic tile, durable and easily cleaned, also offers numerous design possibilities, including compatibility with terrazzo. (*American Olean Tile Company.*)

much also has to do with the type of traffic and area in which this material is applied. For example, in a great many cases it is necessary to use a polymer finish to give proper protection and appearance, but if this is necessary then the expense of the floor is unwarranted, since the claim is that the floor needs no finish. In other cases, the floor has worn to so granular a surface that it soils much too easily. In still other cases the surface has been so smooth as to be too slippery. It is recommended that such floors be considered for only very special applications, such as where bacteria control is a problem, and only then where a reputable and skilled installation contractor is available. (See Fig. 28.)

Floors, Elevators, and Stairs 73

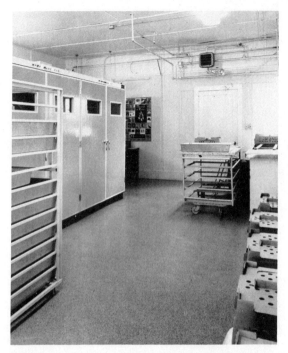

Fig. 28. Plastic laminate floors, such as this clear epoxy resin with ceramic fired-quartz chips, provide a uniform surface free of cracks and crevices. (*H. B. Fuller Company.*)

Plastic Tile

Of considerable interest is the continued development of plastic tile that is literally maintenance-free, a mineral material bonded in an acrylic base. This material is available in a variety of colors and sizes, such as 18 in, 24 in, and 36 in square. It has a thickness of $\frac{1}{8}$ in, and a precoated adhesive backing; preferably it is installed on top of a smooth, steel-troweled concrete subfloor. Rather than butt-jointing, the method by which resilient tiles are typically installed, it comes provided with an approximately 8-in-wide silicone filler between the tiles. The material is quite nonporous, and finishes will not bond to its surface. The only cleaning that is required is wet or dry cleaning for soil removal. The material is an excellent choice for pedestal floors, such as are used in data processing areas. (The wiring beneath pedestal floors is easily damaged by the large amount of moisture typically used in scrubbing and waxing other types of floors.) As production costs continue to decline, and wage rates increase, unquestionably this type of material will see greater and greater use in various types of areas.

A similar material is made for use in restrooms or other areas where ceramic tile is usually found. Installed with a polyurethane grouting, on completion it forms a seamless installation and the floor may be used immediately after application.

Rubber Tile

One of the most expensive resilient flooring materials is rubber tile. Its principal advantage is durability and ease under foot. In most cases, however, another type of material will be a better choice.

Stone

Stone floors—and brick floors as well—provide an extremely durable and handsome appearance (Fig. 29). Stone floors, espe-

Fig. 29. Stone floors, when properly protected from abrasion by sand, make attractive, durable surfaces, even for lobbies.

Fig. 30. Where stone or ceramic is subjected to corrosive materials, an epoxy grout should be used. (*H. B. Fuller Company.*)

cially those having a rough surface, provide good service as soil-entrapment areas. To avoid dusting, and to enhance appearance, stone and brick floors may be successfully sealed, such as with polyurethane or water-emulsion sealers. Good choices for such floors are vestibules, lobbies, patios, lounges, and restaurant areas. (See Fig. 30.)

Terrazzo

Although terrazzo has a rather high initial cost, it is one of the most durable and attractive of all floor materials. If set on a firm foundation and provided with a sufficient number of divider strips, it is virtually indestructible if it is properly maintained with a surface finish. The surface is approximately 70 percent marble chips and 30 percent portland cement and does need a sealer, or finish, to protect its porous structure from staining and soils. When grinding the terrazzo, it is important that the divider strips be ground quite flush, since if they are allowed to project even a small amount above the surface they will provide a constant problem: It is extremely difficult to clean such a surface. To minimize terrazzo floor care, avoid very light colors; preferably, use mixtures of colors.

Travertine

A special type of marble is travertine, a variety found mostly in Italy and which presents a surface that has innumerable cracks and crevices. Although providing an initially rich appearance, it is literally impossible to keep the cracks and crevices clean; attempts to wash or wax such a floor simply leave all the soil in the fissures. Users of travertine flooring material must accept the fact that only the top surface will in any way be clean. Certainly in any type of health care area it should be completely avoided. Synthetic travertine is available as a variety of vinyl flooring material—as if the original were not bad enough! One solution to the problem for both the marble and the synthetic vinyl travertine is to have the cracks filled by the manufacturer with clear epoxy resin. This filling does not detract from the appearance but it does provide a uniform surface that may be easily cleaned. Some bank counters of this material are actually smooth enough to write on. The use of travertine marble for wall covering and other vertical surfaces is much more successful.

Vinyl Asbestos

One of the most successful and widely used of all flooring materials is vinyl-asbestos tile. Typically available in $\frac{1}{8}$ in thickness and in 9- or 12-in squares, it can be obtained in a great variety of colors and patterns. By all means avoid tile that is embossed, grooved, or physically striated; it is literally impossible to maintain these types of tile, no matter what the material. The middle range of colors, such as mottled or striated tan, maroon, and green are easiest to care for; the dark colors tend to show all particles of dust, while the light colors show every scuff mark or rubber burn. (See Fig. 31.) It is especially desirable to choose a pattern that includes a number of black streak marks, as this tends to hide cigarette burns and rubber heel marks.

Fig. 31. The two areas of vinyl-asbestos floor are equally soiled. The proper flooring choice is obvious.

Wood

Despite developments in many types of synthetic flooring materials, wood floors still have a number of advantages. In general, planked floors are to be avoided (except for gymnasiums) as any shrinkage or expansion will give difficulties. On the other hand, parquet floors made of short lengths of wood (typically forming squares of 9- to 12-in size) do not have these disadvantages. Such floors, properly sealed and maintained (with strict avoidance of large quantities of water) will give long service. Some parquet floors have been in continuous use for 40 years or more and still have a rich, attractive appearance. Their principal disadvantages are sound reflectance, and the fact that a scratch or a groove leaves a permanent scar.

ELEVATORS

Elevator lobbies are some of the most heavily trafficked areas in any facility. The most important factor is adequate size to prevent damage to and deterioration of surfaces from overcrowding. Under such conditions, the durability of surfaces is very important, and special attention should be given to floors and walls—as especially discussed both earlier in this chapter and in Chap. 5. Waste receptacles and cigarette urns should be located between every two elevators, on both sides of the elevator lobby.

At least one elevator should serve all floors, including the basement and the roof or penthouse area. Do not omit an elevator even in a two-, three-, or four-storied building. It is necessary for transfer of equipment from floor to floor, moving of furniture, and the like.

Check that the elevators are of the proper size to accommodate the equipment which must be moved from floor to floor, such as carts, automatic scrubbing machines, beds, equipment, desks.

The most practical elevator door is a plastic-laminate material recessed into a stainless steel framing material, facing both into the elevator and out to the elevator lobby. If only one material is to be used, use stainless steel, although scratches tend to show up on such an unbroken surface. Avoid painted, wood, and bronze elevator doors. If, however, doors other than those recommended are provided, then install stainless steel or plastic-laminate kick plates on both sides. Consider the possible advantages of having a rear elevator door at selected levels. On occasion, using an elevator as a section of corridor for the movement of large equipment can save making many turns and causing considerable wall damage.

The ideal elevator floor is carpet, consisting of low, dense, hard pile; looped synthetic fibers, and mixed, tweedy colors. Preferably, the carpet should be rectangular to permit rotation, which will considerably lengthen its life since most of the wear occurs at the entry area. It may be desirable to use a screw-down metal nosing and/or grommets to prevent a tripping hazard or an unsightly appearance. Where carts and other wheeled devices must be used, carpet may not be the best choice if these items are heavy, in which case a ceramic floor with neoprene dividers should be considered. Another good choice would be a synthetic material. The most difficult elevator floor to maintain is resilient tile; even with a great deal of dry mopping, wet mopping, waxing, and buffing, a resilient floor will wear out rather rapidly and normally will be unsightly.

A good alternate to carpeted elevator floors is carpet squares. These are available either in synthetic carpet fiber (such as nylon), mohair, or synthetic textured surface such as polypropylene. The advantage of the tiles, which are usually 12 in square, is that they can be moved, either for cleaning, or to eliminate traffic patterns which will have been formed, and the most badly worn tiles may be simply discarded and replaced.

The ideal elevator wall is made of plastic laminates, such as Micarta or Formica. Railings and grab bars should be stainless steel (aluminum may discolor clothing). Carpet should also be considered for elevator walls; the glued-on types are easily repaired using the "cookie cutter" system. Walls should have hooks at the top in at least one elevator in each building so that protective pads can be hung when moving equipment or cumbersome furniture.

Normally, acoustical control is not necessary for elevators, especially if the floor is carpeted. The ceiling should be simple, such as fluorescent lighting behind hinged plastic panels. Avoid intricate egg-crate louvers or screenings.

A 120-V electric outlet should be provided in each elevator to permit ease of maintenance, both custodial and mechanical. With such an outlet, the interior of an elevator can be cleaned without putting it out of service.

To prevent the elevator floor from becoming a depository for litter and cigarette butts, provide a wall-mounted cigarette urn/waste receptacle combination unit. This provision should be cleared with the local fire marshall and building codes (similar to the application for stairwells).

In installations where a number of errors might be made in elevator floor selection (based on experience factors), or in a building frequented by children, some of whom regularly push the buttons for all floors as a prank, a remedy exists in the design stage: elevator cab controls can include correction buttons to "erase" floors which have been selected but at which the rider does not wish to stop. Under these conditions, the use of correction buttons will eliminate a large enough percentage of stops and starts to have an effect on maintenance and operating costs, and certainly on complaints and public relations.

Having a large freight elevator available can save innumerable minor—and major—damages to walls, doors, and stairs caused by moving furniture and equipment. It might also avoid some nasty accidents. Speed is not important, and consideration should be given to a hydraulic elevator for freight handling where only a few floors are involved. A freight elevator should serve all levels, including the roof.

Try to place the freight elevator so that it is easily accessible from the outside of the building, thus avoiding the need to bring heavy or bulky items through the building and possibly damaging it. Specifically, try to avoid the necessity of moving furniture, equipment, garbage, or the like over finished floors such as carpeted areas or resilient tile. The freight elevator might be next to the loading dock, with a door opening onto the dock as well as an opposite door opening to the inside of the building. A holding or receiving area adjacent to elevators is important in order to permit the use of the elevator at the proper time, without crowding or damage.

Each elevator shaft should contain lights for maintenance, inspection, and cleaning. Elevator-shaft walls should be as smooth as possible to prevent accumulation of lint and grease. Shaft pits should be well lighted and should contain a convenience electric outlet. The pit floor should be smooth, troweled concrete to provide for ease of cleaning.

The elevator control room requires enough space to both hold and maintain the proper equipment. The room should be well lighted and ventilated, and the floor area kept free of electric conduits.

STAIRS

The poorest material for stair treads and landings is resilient tile, such as asphalt tile and vinyl asbestos. It wears quickly and is extremely difficult to keep clean and finished. A successful material for the treads and risers is precast and ground terrazzo, with carborundum nosing strips properly recessed. If the landing

Fig. 32. Where carpet is to be used on stairs, a molded nosing should be used to avoid excessive carpet wear.

is terrazzo, however, it will require considerable care, so it is better to make the landing of another material, such as carpet, especially if the carpet can be removed for cleaning elsewhere. Such a use of carpet provides good soil entrapment, similar to that used in an elevator, and eliminates soil transfer from one floor to another. (See Fig. 32.) Another good tread and landing material is formed rubber, which looks very much like a waffle-grid pattern. The material wears well and keeps a good appearance without much maintenance, especially in the brick-red coloration, but of course periodic dry and wet cleaning is required, although finishing is not necessary. The use of carpet for the stair treads themselves is not a good investment, and typically should be reserved for executive or highest-quality situations only. When carpet is used on treads, it should be laid in such a way that the pile rows are diagonal to the line of the tread, avoiding opening up the pile as it bends over the edge of the tread, which leads to very rapid wear; or better, provide a nosing of another material. Avoid bluestone, slate,

Fig. 33. White-on-white terrazzo is not suitable as a stairway material.

and marble, since these are soft materials and quickly develop cuplike wear spots that are hazardous. (See also Fig. 33.)

The inside of the stairwells should be finished in such a way that they do not require periodic redecorating or excessive attention. Handrails should be made of stainless steel or aluminum; wood requires too much attention, brass requires polishing, steel requires painting, and some other materials are either too fragile or too expensive. Stairwell walls should be either plaster with a gloss enamel coating for easy washing, ceramic or epoxy-coated where there is a great deal of traffic and potential for graffiti (such as in a public school), or covered with a heavy-duty vinyl wall covering. Stairwells with a natural brick finish are also practical and give the architect an opportunity to bring the outside "inside."

Handrails should be firmly anchored so that they will not work loose. Railing posts should not be anchored in the treads, as they are difficult to clean around, but in the sides of the treads or, where possible, in the walls. The design should not permit the use of the railings as a ladder for children to climb, or for adults to sit on; further, their design should prevent catching one's finger in a space of diminishing size and should be free of sharp edges and corners. Handrails should be continuous where possible. Where wrought-iron railings are to be used in stairs, balconies, or other areas, they should be provided with a plastic or vinyl coating to eliminate finger markings and rust and to simplify cleaning.

Do not allow any space between the stringers of stair runs; the vertical shaft that this creates permits objects to be dropped down it.

Lights should not be located over the stairs themselves, as then their replacement and cleaning becomes hazardous. If lights must be located in the ceiling, then they should be over the landing, but preferably the lights should be in wall-mounted fixtures which can be reached from the floor for cleaning or relamping.

A wall-mounted combination cigarette urn and waste receptacle (or two separate devices) should be installed on every stair landing to prevent the landing being used as a refuse collector. The use of cigarette urns and waste receptacles in stair landings should be approved, as to both type and location, by the local fire marshal and any building codes that are applicable.

A standard electric receptacle should be installed at every stair landing to make adequate maintenance possible. The outlet will be used for floor-cleaning equipment, wet or dry vacuuming, wall washing, hand drills, and so forth.

Do not use stairs of less than three risers; use a ramp instead, with a gradient not exceeding 1 ft in 12. Otherwise, provide a ramp alongside the stairs for equipment and furniture moving, and for the use of wheelchair workers or visitors. Heavy safety-tread edges should be provided, typically using carborundum strips or impregnation.

CHAPTER FIVE

Walls and Ceilings

The higher one goes above the floor, the less important maintenance, and therefore design for maintenance, becomes. In order of importance one would increase from ceilings to upper walls to lower walls to floors. But this is not to say that walls and ceilings are not an important aspect of this subject and deserving of serious consideration.

WALL COVERINGS

Vinyl wall coverings can be an excellent investment as they eliminate painting and their cleaning is very simple, with two big *ifs*. First, a pattern must be selected in a heavier-weight vinyl that is not too deeply or sharply embossed (Fig. 34). If the embossing contains a number of sharp crevices and cracks (such as in some grass-paper patterns), then soil will lodge in these crevices, and attempts to wash it off will result in embedded soil, a very unsightly appearance, and a great deal of lost time. Therefore, a pattern should be selected with an embossing consisting of smooth curves. Second, installation is very important, and any peeling or gaps left between sheets of the film will create difficulty. Further, if the wall surface is not prop-

Fig. 34. This mottled-pattern fabric-backed vinyl wall covering will hide stains and soils, yet is easily cleanable. (*Multicolor Company, Div. National Gypsum Company.*)

erly prepared, bleeding of mastic or other chemicals can ruin the vinyl surface. A small extra quantity of the vinyl should be bought initially to repair tears and other damage.

The use of wallpaper is practical principally for an area which involves a low density of traffic. Plastic-coated paper provides the most durable, cleanable surfaces. Foil papers are easily cleanable but are also easily torn or scratched. Natural-grass papers present a combination of continual problems. Flocked papers require regular vacuuming to avoid a dust build-up.

Avoid such impractical wall coverings as burlap, natural-grass paper, and felt. Epoxy wall coatings provide a smooth, continuous surface that gives good light reflectance, colorful appearance, ease of cleaning, and a surface that is difficult to write on. (See

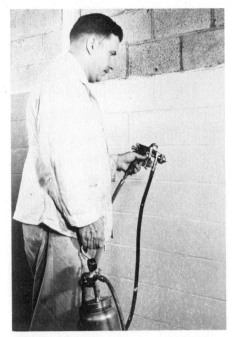

Fig. 35. An epoxy finish converts a concrete block wall into an easily maintained surface that resists graffiti and vandalism. (*McDougall-Butler Company.*)

Fig. 35.) Installations of concrete block, often chosen for a very low initial cost, have resulted in extremely unsightly conditions after just a few months of use. A good solution to this problem, either at the time of initial construction or later on if necessary, is coating with an epoxy resin at least up to a 6-ft height. Actually, an epoxy-coated wall provides a more sanitary, cleanable surface than a glazed-block wall because of its continuous nature. Caution: Do not put epoxy coatings on dry wall since when such a wall is punctured the repair, patching, and refinishing is very troublesome. The epoxy treatment as described is most satisfactory on plaster or concrete block. Further, the standard or lighter colors are more durable than the darker "decorator colors."

Gloss enamel is the only paint that combines durability, good light reflectance, and a tough surface, with complete washability; semigloss enamel has these characteristics to a more limited extent. On the other hand, the flat paints, especially the latex varieties, although easy to apply and initially good economically, mark easily and are difficult to clean. Typically the latex-painted wall is repainted much more frequently than the enamel-painted wall. On the other hand, feature strips and dado color breaks are to be avoided. For walls, and other surfaces as well, avoid paints that have been mixed to order, as the duplication of the color may become difficult or time-consuming. Standard paints are now available in such a tremendous variety of shades that they should be suitable for all but the most extreme conditions.

Many interior decorators have thoughtfully decided that art reproductions provide a colorful and beneficial value in such buildings as hospitals, schools, hotels, colleges, nursing homes, public buildings, and the like. The problems associated with frames, their repair, painting, and cleaning, are completely avoided when a reproduction is enclosed in plastic. In this case, not only can the surface be quickly and safely cleaned with a dusting cloth or a damp cloth, but the reproduction can also be easily handled without damage (removed for wall washing or transferred to another location).

In rooms where wall hangings are frequently used and changed, such as in a conference room, sales display room, or art studio, it is desirable to place a strip—on which hangings can be affixed—along the full length of at least one wall, approximately 78 in above the floor. Where tapes will be used to attach the hangings, a stainless steel strip about 2 in wide is ideal; where thumbtacks are to be used, a cork, rubber, or similar strip should be provided. Ideally, a strip combining these characterisitcs would cover all contingencies.

If lighted paintings are to be hung, such as in a boardroom, an outlet should be provided for each at a 60-in height, and all these lights should be controlled from a single switch.

Provide for the location of bulletin boards in service areas, but avoid using them in traffic areas.

WALL MATERIALS

Glazed tile provides an extremely high quality of wall surface, although its first cost limits it to such specialized areas as food processing, pharmaceuticals manufacture, certain hospital areas, and the like (Fig. 36). It should be borne in mind that the cement grout between the tile is of an entirely different, porous structure, and, to meet the high conditions of the tile itself, it should be sealed properly, such as with an epoxy resin—or the tile itself should be set in an epoxy grout.

Although typically considered to be a luxury material, wood paneling can actually be a low-maintenance surface. This is especially true when the wood has been treated with a baked-on finish or plastic surface. It is of great importance to equip furniture so that the paneling is not scratched by the moving of chairs. Where a good deal of equipment or furniture movement is anticipated in a paneled area, a dado at the proper height should be installed. Where the appearance is acceptable, the

Fig. 36. Ceramic wall surfaces resist damage and soiling, yet are easily cleaned. (*Abreu & Robeson, Architects, Atlanta.*)

Fig. 37. Fused plastic finish on tempered hardboard provides the advantages of marble without its porosity and softness. (*Masonite Corporation.*)

plastic-covered woods are preferable because they can be wiped clean with a damp cloth and they resist scratching and other damage. (See Fig. 37.) Fading of wood paneling will become noticeable if it is not protected from direct sunlight.

The maintenance cost of plastered and gypsum wallboard walls, other than for a very lightly trafficked and used area, is high. *From the surface standpoint, regular painting and washing is necessary.* The wall is easily damaged mechanically, and a good deal of time may be required to patch grooves, scratches, and broken corners. The very act of this patching, involving wet patching material, and then, after a time, repainting, introduces opportunities for further damage and soiling. Where possible, then, more durable surfaces should be used, such as plastic, stone, ceramic tile, or brick. Concrete block may also be used, but should be coated with epoxy resin to prevent marking and soiling.

Fig. 38. Baked enamel on steel can be a practical wall surface in dormitories. (*H. H. Robertson Company.*)

Steel-enameled walls provide durable yet cleanable surfaces in selected locations. (See Fig. 38.)

Cracks in plaster can be avoided if care is taken to locate casing beads at terminating points such as against windows and door jambs, using back-to-back casing beads for control joints, and using such beads wherever there is a change in the substrate material. Doorframes should be of the wrap-around type so that cracking does not occur next to the frames.

Plastered wall corners provide the greatest resistance to damage where the plaster is finished to a quarter-round metal corner rather than to a relatively sharp corner angle bead.

Gymnasium walls receive a lot of abuse, and are typically covered with handprints—and footprints—from athletes and others. These walls should be protected by impervious materials, such as epoxy resin or ceramic tile, up to the height of the doorframe. Where walls are to receive the installation of net or game ties, provide solid, heavy, anchor backing.

Do not use gypsum wallboard that is less than $5/8$ in thick, as it would be too susceptible to damage. Preferred are two $3/8$-in or two $1/2$-in panels glued together with overlapping joints.

There are numerous considerations when it comes to the use of concrete block for walls. Here are some of them:

    1. Raw concrete block, of course, is the least expensive approach, but this very porous, grainy structure deteriorates almost before your eyes, becoming dirtier and dingier, and is very easily and permanently damaged by all types of marking instruments.

    2. Painted concrete block is an improvement, but the right type of paint must be applied carefully to secure adhesion. For high-quality appearance, frequent repainting, as well as spotting and washing, is necessary.

    3. Much more durable and attractive than painting is the coating of the entire concrete-block wall, including grout lines, with an epoxy resin. The resulting surface is almost glazed in appearance and can be made quite colorful. It is easy to clean and difficult to mark.

    4. Glazed block can be purchased, but this leaves the question of the grout lines. Epoxy grout helps to solve this problem.

    5. Where the design includes an outside corner, such as where a corridor turns, rounded blocks should be used, because sharp concrete-block angles are easily damaged.

    6. Be sure that a molded cove base is applied so that the bottom portion of the concrete-block wall does not become soiled through the use of mops, floor machines, and so on (Fig. 39). If a halfblock (4 in high) is used on the bottom course, then a 4-in cove base will overlap the first grout line, leaving a groove which collects dirt that is almost impossible to remove. In such cases, the half-block should either be placed at the top of the wall, or a 6-in cove base should be installed.

Attempts to face elevator-cab interiors with metal walls have been largely unsuccessful, as these are easily damaged by scratching, or even by being written on with certain types of

92 Building Design for Maintainability

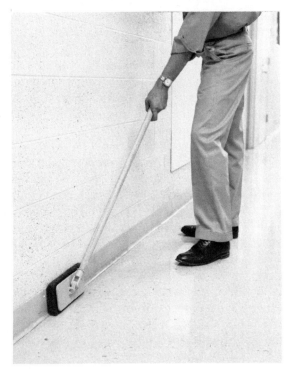

Fig. 39. The lower portion of concrete block walls will become black with soil unless protected by a cove base. (*3-M Company.*)

marking pens. The most successful elevator wall—resistant to most damage and easy to clean—is the high-pressure laminated-plastic material such as Formica or Micarta—and fortunately these materials are relatively inexpensive in comparison with metal. Carpet for wall surfacing in elevators is also being used successfully, especially where the walls are subject to damage from carts, and so on.

Cove bases of vinyl or rubber are an absolute must; wood bases should be avoided. Where walls are installed without bases (such as paneled walls or marble), invariably a soil line will form on the lower few inches of the wall—caused by mop and floor-machine marks. Even the rubbing along the wall of a treated dusting mop will leave such marks. Actually, cove bases should be considered a part of the floor as they are properly installed at the time of floor installation and also cleaned while the floors are cleaned. Cove bases should not receive finishes or sealers, however. Carpet baseboards made of the same material as the flooring itself should be considered for carpeted areas. Ideally, baseboards should be recessed, with cove (rounded) corners. Baseboards should be 6 in high in corridors, 4 in high in other areas, and provided with a backing in the wall to prevent wall damage from floor machines, carts, and other devices, if the wall is easily damaged (as thin wallboard).

Because lobbies are subjected to so much traffic of all kinds, special attention should be given to the wall treatment. Polished stone, granite, or marble, provides ease of maintenance with resistance to staining and damage; wall surfaces here should extend to ceiling height if possible.

DESIGN FOR WALLS

Writing on walls is simply human nature—excavations of the Egyptian pyramids have disclosed graffiti! And the question is not one of changing human nature but of establishing conditions that discourage this activity. Graffiti cannot be eliminated completely, but it can certainly be controlled and reduced to a minimum amount. While much can be done in the operational phase (rapid removal, prosecution of offenders, and the like) still more can be done in the design stage. Of greatest importance is illumination; graffiti is sharply reduced where illumination is heightened—as, for example, when the illumination in a restroom is increased from 30 fc to 100 fc. Further benefits are achieved through the use of bright and colorful paints and wall coverings, as well as materials that are difficult to write on, such as plastic laminates, glazed ceramic tile, epoxy coatings. By the way, do not waste money on signs or posters. They simply do not get results.

Insofar as possible, avoid indentations, projections, crevices, recesses, pilasters, and other wall features that break up the smooth surface. Not only is additional area created which must

Fig. 40. Unless walls are continuous, indentations will harbor soils and require hand cleaning.

be maintained and painted, and many more corners produced which may become damaged, but such a wall may invalidate the use of a wall washing machine and require manual cleaning, as shown in Fig. 40.

For heavily trafficked areas, provide a minimum width of 8 ft for corridors in order to avoid wall damage. Where passageways or other areas are subject to a good deal of traffic by vehicles and carts, normally a good deal of scuffing, scraping, and gouging of the walls is seen. This is best avoided by providing either a concrete wheel curb or a guardrail to protect the walls from damage (see Fig. 41). Consider rounded corners in heavily trafficked intersections. This also provides for smoother flow of traffic, less wall damage where carts are used, and improved safety.

In a steel-frame building, use of prefabricated walls eliminates the cracking induced in solid masonry walls when thermal changes or other shock cause structural movement.

The consideration of movable partitions is principally a question of how much change is anticipated. The cost usually balances out by the time the partition has been moved three times. A general approach to solving this problem is in the "office-land-

Fig. 41. In areas involving vehicular traffic, columns require corner protection. (*Pawling Rubber Company.*)

scaping" concept, which permits a permanent location or regular movement at only nominal cost.

Where a good deal of furniture movement and materials handling is expected, to prevent excessive scoring and damage consider the use of steel or stainless steel corner guards for walls and columns (Fig. 42). The guards themselves should have rounded corners rather than sharp corners. Practical formed-plastic guards are now also being marketed.

Fire extinguishers should be enclosed in recessed cabinets with flush, easily openable fronts.

ACOUSTICAL CEILINGS

The purpose of acoustical ceilings, by the definition of the name, is sound control. Optimum selection would be of a material that provides a uniform level of sound control throughout the life of this ceiling, as well as good appearance without an undue expenditure of time in its maintenance. Unfortunately, most acoustical ceilings do not fall into this desirable category. There are a few examples of acoustical ceilings that can be cleaned

Fig. 42. The rails protect walls from damage by carts and wheelchairs, while also providing a significant safety factor for hospital patients. (*Abreu & Robeson, Architects, Atlanta.*)

successfully, however. Other varieties are easily cleaned but have little or no acoustical value. A good rule is to make all acoustical ceilings removable. And for best sound control, the partition walls should extend through the ceiling to the underside of the structure above.

Sprayed-on mineral ceilings are typically of either the soft or hard variety. The softer the material, the better the acoustical quality, but the more difficult the maintenance (Fig. 43). The softer material is very easily damaged by scraping or scratching, and these all-too-visible marks are impossible to remove. The soft-mineral material has earned the questionable distinction of being one of the very few materials of any kind that is absolutely noncleanable; that is, it cannot be vacuumed (the bristles of the vacuum brush dislodge particles which rain down like snow), washed, or cleaned in any other way. The harder varieties may be very carefully vacuumed or damp-cleaned, but this is time-

Walls and Ceilings 97

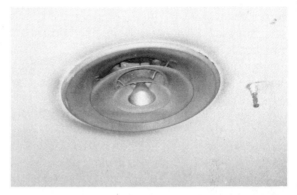

Fig. 43. Soft-blown mineral acoustic ceilings become quickly soiled and easily marked, yet are very difficult to maintain in any way.

consuming. Both varieties may be spray-painted to improve their appearance, but even with the special latex paints made for this purpose the acoustical quality is reduced with each application. Although they provide an inexpensive first application, blown-mineral ceilings take an immediate downhill direction in terms of appearance and acoustical quality.

Many acoustical ceilings are made of small or large rectangles of wood or mineral fiber; the mineral fibers have the advantage of being fireproof and should be used exclusively. These materials may be directly attached to ceiling surfaces, or suspended in pan fashion. The mineral materials—especially the harder types—may be vacuumed and damp-cleaned to a certain extent, although they are never completely cleaned in these ways. The typical user will repaint the ceilings from time to time, with the resulting loss of acoustical value. In addition, the perforated materials will tend to have their smaller holes filled with paint, leaving an unsightly appearance. Such materials should be used principally for economy installations. To ensure the availability of repair and replacement materials when needed, specify standard popular patterns on any such ceiling materials that the manufacturer will keep in production or have available on special order.

Where acoustical tile is used, poorest results are obtained by gluing or nailing the tile to the surface above. Best results are obtained by a suspended ceiling, which also provides an intermediate space for lighting, wiring, ventilation, installation, and the like. Of these, there are two varieties. The splined, or clip-type, ceilings fits each tile into the metal support members; to remove one individual tile it may be necessary to remove an entire row or area of them. A much better system is the lay-in ceiling, where the tiles are supported by individual rectangular frames. Thus, each tile can be lifted out individually and the area above it exposed. Typically, lay-in ceilings provide for panels that are 2 ft square or 2×4 ft, the latter being very desirable for modular construction and the installation of flush fluorescent-light fixtures. Avoid wood-fiber acoustical backing, as it presents a fire hazard and is vulnerable to infestation by pests. Avoid the installation of acoustical tile in toilet rooms, shower areas, kitchen areas, stairwells, and on exterior surfaces. If acoustical treatment is required for swimming pool or hydrotherapy areas, the use of perforated aluminum pans (Fig. 44) or of plastic-coated fiber-glass baffle panels will be successful. Where acoustical ceilings are not of the lift-out variety, provide metal access doors or access tile systems to simplify maintenance of valves, regulators, steam traps, junction boxes, and so on.

One of the most successful types of acoustical ceilings are the mineral-fiber materials which are faced with a polyester or mylar film. Although the acoustical quality is somewhat diminished, yet it is certainly acceptable as an acoustical ceiling and the film surface makes it possible to literally wash the ceilings an indefinite number of times (although care must be taken to avoid abrasive materials, such as scouring powders, which will tear the film). Naturally, the initial cost is higher than for unfaced mineral-fiber ceilings, but this is an excellent material where high sanitation is desired, such as in food-service areas, executive suites, and the like.

In addition to mylar and other plastic films over a smooth fiber glass ceiling tile, it is also possible to obtain plastic-coated (perhaps vinyl) mineral-ceiling tiles, as in the fissured pattern, for example. Such a surface decreases soil retention, simplifies

Walls and Ceilings   99

Fig. 44. Perforated aluminum-pan ceilings are long-lived, acoustical, and may be washed repeatedly. (*Gold Bond Building Products, Div. National Gypsum Company.*)

cleaning, and may avoid the painting of ceiling tiles that would otherwise be required. (See Fig. 45.)

Adequate air relief should be provided to prevent the dislodging of lay-in tile by sudden air pressure from below. This is especially a problem in foyers and entranceways.

Metal sheets may be perforated and backed with mineral (such as glass wool) materials to provide acoustical surfaces that may be washed and that resist physical damage. By all means use aluminum pans (Fig. 44) rather than steel pans to eliminate rusting. And in the cleaning procedure avoid excessive water, as it may be forced into the holes and the mineral backing material and then drip back out carrying soil. Although the initial cost is rather high, the material is easily washed and

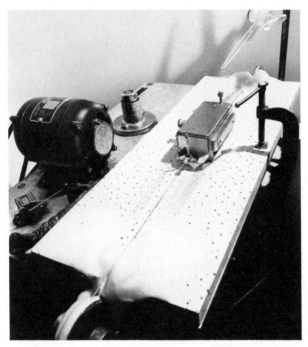

Fig. 45. Acrylic-coated acoustic tile withstands repeated washings, as demonstrated in this laboratory test. (*Gold Bond Building Products, Div. National Gypsum Company.*)

repainted where desired. Excellent installations are locker rooms, food-service areas, health-care areas, and the like. It should be remarked, however, that many organizations find metal-pan ceilings are too "institutional" in appearance.

Modern construction often calls for a ceiling system of modular dimensions—often 4 or 5 ft square—incorporating lighting, acoustical control, heating, ventilating, and air conditioning into its integrated design. Such modules typically take a pyramidal shape, with the anemostat or air diffusor at the apex. Other integrated ceilings are perfectly flat, with the whole perforated surface acting as the diffusor. The surfaces may be of mineral-fiber, metal-pan, or membrane-faced materials, incorporating the advantages of cleanability in the latter two cases. The design of such ceilings often eliminates the formation of black rings around the diffusors—a very desirable feature. Such integrated ceilings—where multitrade maintenance personnel may be utilized or mixed trade personnel used in a team—can permit cleaning, relamping, and electrical inspection or repair, all in a single activity.

Thus, ideally, by using the space between a drop ceiling and the bottom of the floor above it as a plenum, and utilizing a perforated metal-pan ceiling so that much of the entire surface is used for the diffusion of air, it is possible to completely eliminate anemostats, diffusors, and grills, and thus to eliminate their cleaning and painting; this also eliminates the formation of black soil deposits typically seen around ceiling diffusors, which often create the most unsightly aspect of a building area. The soil deposits which appear around each small opening in the plenum are less noticeable and more easily handled.

Another approach to lessen the amount of soiling usually found in circular patterns around air diffusors is the installation of a special type of surface immediately next to the diffusor. This might be a 6-foot square area of smooth acoustical tile (rather than fissured) so that it can be much more easily cleaned; or it might be a circular metal surface projecting a foot or more from the diffusor itself and constructed of the same material, which can then be washed along with the diffusor.

CEILING DESIGN

Corridor ceilings are often lower than those in other areas because of piping and ducts which are concealed above them. If this results in the ceiling being too low, damage and vandalism will result. The damage may be caused by the movement of ladders and equipment or poking with the end of a mop handle. The vandalism will be in the form of handprints (even footprints!), graffiti, and tiles displaced, damaged, or removed. If such problems are anticipated, an increase in the distance between floors should be considered. Also consider the possible provision of a wiring raceway for telephone, intercom, and other wiring (such as instructional media, music systems, synchronized clocks, and alarms) to minimize the necessity to remove and replace tile, which more often than not becomes soiled or damaged in the process.

Some more modern designs provide for complete ceilings of plastic, with lighting behind. The plastic may be solid or open. The open type of material gives great difficulties in dust removal, whereas the solid type tends to collect refuse on the inner surface where it is easily visible and rather unsightly. In either case, the plastic ceilings should be removable in order to service the electrical and other equipment behind them, and also for vacuuming and washing. It is important that spare plastic pans be obtained at the time of installation, as replacement will be needed both for cases of natural cracking or crazing and for other unavoidable damage (Fig. 46). A 10 percent overage would be

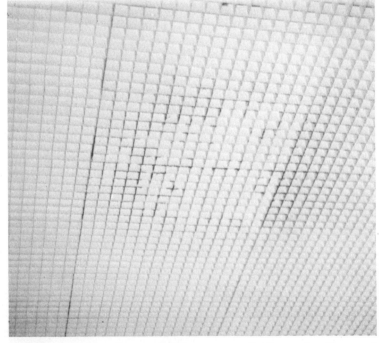

Fig. 46. Egg-crate ceilings and diffusors are easily damaged, collect much soil, and are difficult to clean.

a reasonable number, and these should be rotated into the regular material periodically to retain a consistency of coloration and appearance.

A smooth, plastered ceiling, with a good-quality painted surface, provides a minimum maintenance demand and is easy to clean. The lack of acoustical qualities may be overcome by using carpeted floors, drapery, and other sound-absorbing materials. A sand finish, however, although originally attractive, is much more difficult to clean and makes any patching immediately apparent as it is difficult to match the old surface.

Gypsum wallboard material ("dry wall") is very widely used in ceiling construction, especially in residential buildings. The construction is quick, clean, and economical. The laminated construction (two $3/8$-in sheets glued together with overlapping joints) tends to eliminate all cracking as compared to the single sheet ($5/8$-in thick) construction, while also being stronger and less apt to transmit sound. The skill of the installers will often make the difference concerning later appearance in terms of protruding nails, smoothness of joints, and the like. As with plastered ceilings, the quality of the painted surface is very important, and a sand finish should be avoided. A textured finish on dry wall has a tendency to flake off and is also very difficult to wash.

In industrial and commercial installations, the visible ceiling typically amounts to a collection of pipes, ducts, wiring, and the like. To save a great deal of cleaning time, from the standpoint of appearance, some organizations will "paint out" the ceiling through the use of some flat, light-absorbing paint. For example, all ceiling surfaces will be painted a dull color (such as frosty plum) down to the level of the light fixtures. In such a case, especially where the lighting is of high intensity, the upper surfaces tend to disappear. On the other hand, some organizations will color code all piping and duct work, including the application of pressure-sensitive labels indicating the type of material in the piping and the direction of flow. This approach is not only utilitarian but can also be quite attractive. Special vacuum attachments, such as curved bristle tools for the tops of pipes, are available to permit cleaning from the floor without scaffolding or special equipment, except for very high-ceiling areas. In those industries where oily types of materials become deposited on surfaces, or where there is a good deal of lint or paper dust in the air, such ceiling maintenance becomes mandatory from a fire-prevention standpoint. In other cases, ceiling cleaning is necessary to provide quality control where machine vibration would cause dusty particles to "rain" down.

When certain types of equipment are installed above drop ceilings, such as pumps or piping, any leakage or failure of this equipment is likely to damage the ceiling below. Also, such equipment may be difficult to reach or work on without damage to the grid system which holds the ceiling panels, and during such maintenance ceiling panels are typically badly soiled or damaged. The use of skylights should be avoided, as their maintenance costs are expensive. Extensive damage to floors and equipment may be caused through the simple leakage of one point in a skylight. Substitute artificial lighting where necessary.

CHAPTER SIX

Furniture, Fixtures, and Fenestration

Somehow, furniture, fixtures, doors, and windows seem to be especially susceptible to problems in maintainability design. Perhaps they end up near the bottom of the list in terms of the application of funds so that by the time more money than anticipated has been allocated to site preparation, foundation, structure, and exterior (typically because of inflation and construction delays), they appear to be logical places to make "cuts" because of their seemingly relative unimportance. Yet, as we shall see, improper consideration of maintainability in these areas can lose the opportunity of a great savings, annually repeated, or conversely cause an expenditure far in excess of that which should have been experienced.

Fortunately, changing styles in architecture and interior decoration have simplified some aspects of maintenance. Just as we see less exterior "gingerbread" such as belfries, columns, parapets, dormers, cupolas, fake chimneys, arches, gargoyles, and friezes, so do we see fewer furnishings in the Victorian manner or trim in the baroque or rococo manner. Although simplicity is the trend, there are periodic lapses, such as in intricate screening or grillwork, that cause much difficulty.

The layout of office (and other) areas will have much to

do with the time required for cleaning these areas, not to speak of the efficiency of general work performance. Floor-surface moldings to cover electric wires are disastrous from the standpoint of floor care; where these moldings are in use it is simply impossible to provide reasonable floor care even with tremendous expenditures of time. Desk and furniture arrangements should be made so that straight, uncluttered aisles are provided, preferably of a minimum 36-in width. Cleaning becomes extremely difficult in "bull pen" areas where as many desks as possible are jammed into a given area; at least 50 ft^2 per person should be provided as a minimum. To afford privacy, rather than using numerous partitions and screens, consideration should be given to sound conditioning by the use of carpeting, drapery, and planters.

In buildings which use standard aisles and corridors, these should be "straightened out," since bends, offsets, or turns create corners which may become damaged, congest traffic, represent safety hazards, and typically eliminate the possibility of the use of an automatic scrubbing machine or other large pieces of floor-care equipment, as such equipment is only of value in open areas or straight corridors. Similarly, curved corridors (such as in circular buildings) often require the design of specialized equipment for their maintenance and pose significant problems in security since only a small length of the corridor can be seen at any one time.

Not only the Quickborner concept, but other office layout systems make considerable use of plantings, both to enhance appearance and to control sound. Unquestionably synthetic plantings are the best investment economically as they have an indefinite life and only an occasional cleaning is required. Natural plants should be used only in interior areas where specifically required by an owner who has been made aware of the attendant problems. These problems can include damage to carpets and other types of floor areas, and to furniture, by overwatering or through leaky planter boxes. Where a building contains a number of natural plants or growing botanical exhibits, a service room for storing tools and supplies of ample size should be provided near other service facilities; or at least a specific portion of a larger area should be so designated.

Where a building may be used fully at one time and then only partly at another (such as a public school that would be only partly in use for evening classes), it is useful to have divider doors that lock off the unused portion of the building. This prevents the use of certain restrooms, corridors, and other facilities during that time, and thus avoids their double cleaning. If such a general condition is expected but the details are not yet known, it should be possible at least to lock off half the building (and still have suitable safety exits).

The subject of smoking regulations should be discussed with the building owner during the initial design stage. In some religiously oriented organizations, including some universities, smoking is absolutely forbidden, thus eliminating the need for cigarette urns and other control devices. Even some businesses have such a regulation. Of course, regulations do not prohibit smoking unless the building owner has specifically determined that this will be done; for example, many teachers smoke in classrooms while standing underneath a sign which proclaims in bold letters that smoking is absolutely forbidden by order of the fire marshal. Some hospitals and nursing homes are refusing to install cigarette vending machines, or to sell cigarettes at their public shops, on the basis of their health hazard; others go further and prohibit smoking entirely within the facility. The management's intent with regard to smoking, and what it actually thinks it possible to achieve, will not only bear on such things as cigarette urns, but also on the types of furniture, flooring, and other surfaces selected, since some will stand up to cigarette smoking where others will not. (See Figs. 47 and 48.)

In facilities like hotels, motels, hospitals, country clubs, and others where rooms become available for cleaning and maintenance at unpredictable times, a great deal of time can be saved by the installation of electronic room-notification systems. On the one hand, these notify maintenance personnel when the occupant has departed so that the room is available for maintenance activities. On the other hand, it permits the maintenance person-

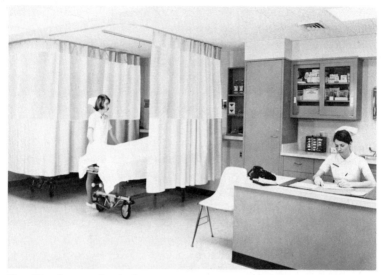

Fig. 47. Plasticised cubicle curtains are flameproof, provide for control of vision, and can be spot-cleaned in place. (*Tami Products Company.*)

Fig. 48. A column treatment providing desirable features: double cigarette urn and waste disposal fixtures, corner protection, electric outlet, carpet base.

nel to notify the business office or the desk clerk when the room has been made ready for the next occupant. Without such systems a vast number of phone calls and other communications are required. As with other evaluations, consideration of such a system should be based on many factors, certainly including in this case the interest on the money invested in the system versus time saved based on average anticipated wage rates and related costs.

Also, to aid in cleaning, provide a cloak closet for visitors next to each bank of elevators so that in bad weather dripping water from raincoats and umbrellas, and trackage from overshoes, will be limited to a small area. Be sure to provide ventilation to these cloak closets to keep them dry and to remove odors. Where theft could be a problem, provide lockable hangers or lockers.

WASTE COLLECTION AND DISPOSAL

The design of the building should accommodate the type of waste collection and disposal system and equipment that will be used (Fig. 49). Consider the space and electrical requirements for can- and bottle-crushing equipment (Fig. 50). Garbage storage areas should be equipped with water outlets and steam outlets and a floor drain.

The collection and storage area for trash or garbage should be located for accessibility so that the time required to bring refuse to this area will not be excessive. The location should not interfere with other activities, such as shipping and receiving. If the central collection area or container is at the general loading dock, it should be to one side of the dock, out of the line of traffic so that soils and litter will not be tracked into the building or onto a freight elevator. Typically, a large container or compactor would be located below the dock, on the paved trucking area, so that materials can be dumped into the container from the dock without strain or hazard or the necessity to use more than one worker.

A number of organizations are required to identify containers of collected waste, date them, and put them in storage, say for one month, so that any losses through inadvertent discard can be recovered. Examples of this are banks, certain insurance departments, research facilities, military agencies. If there is such a requirement, provide a suitable storage facility next to the final disposal area. Actually, it is desirable to provide two such

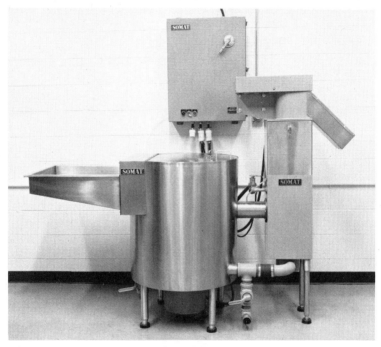

Fig. 49. Building design should anticipate the waste disposal system to be used. (*Somat Corporation.*)

areas, for example, so that one alternates between the room being filled and the room being emptied. Fire protection is also an important factor in these locations.

In locating waste receptacles, include as a factor in each location the disposal of waste if there may be a considerable quantity of it. This waste might be in the form of data processing cards, paper, carbon sheets, chemicals in many forms, chips, and reject materials.

Fig. 50. Some waste control systems, such as this ultrahigh compactor, provide unique solutions to problems but may require considerable space and electric capacity. (*Electronic Assistance Corporation.*)

VENDING MACHINES

Vending machines of various types scattered about in odd areas of buildings cause considerable difficulty in terms of floor damage through leakage, floor spotting as people return to their work areas with beverages, burned floors from cigarette butts, utilities problems, traffic congestion, and problems in refilling the machines and removing their waste. In a campus type of environment, such as a research facility, smaller college (or a natural subdivision of a larger educational institution), a relatively small office park, and the like, the problem can be overcome through a central vending facility. If the buildings are arranged more or less in a quadrangle or circle, this facility would be located at the hub, or center, at the intersection of the walkways that connect the various buildings. The vending structure itself would be encircled by a broad walkway. Such a building could contain not only a wider variety and greater number of vending machines than might be found spotted about elsewhere, but also lounges and snack areas where people might enjoy their coffee breaks or snacks. The structure should contain telephones, restrooms, and drinking fountains, and would also be ideally located for the central security office and possibly central communications as well.

Where a central structure is not provided, consider the provision of a central vending room in each building, which might be located near a loading dock for ease in servicing. The advantages of a central vending room include:

1. Convenience for most users
2. Ease of supervision to minimize security problems, noise, graffiti, and vandalism
3. A minimum of maintenance
4. A minimum of service time by vendors

At the least, avoid placing vending machines immediately against the wall, which makes it impossible to clean behind them.

FURNITURE

Although furniture may not be part of the initial cost of construction—though in many cases it is, such as with built-in furniture—it is still a major consideration before occupancy and during renovation or relocation. It does not have the life expectancy, typically, of the building itself, but nevertheless it lasts a good number of years and represents a considerable investment—and surely is an important consideration in maintainability.

Even though many people normally think of wall-mounted bookshelves and other simple devices as being reasonable, they may forget that almost any piece of major room furniture can be wall-mounted. (See Fig. 51.) With such a design, the floor is completely clear for floor care (except for chairs and wastebaskets, which can be placed on top of the other furniture).

Furniture, Fixtures, and Fenestration   113

Fig. 51. Cantilevered furniture allows rapid litter removal and carpet care. (*Reynolds, Smith & Hills Architects, Jacksonville.*)

Further, where furniture is recessed, only the front surface requires attention, as the sides and top are not exposed. Examples of this are file cabinets, bookshelves, cabinets, fire extinguishers and hose boxes, and display cases.

Furniture surfaces should be cleanable and should resist marking and scratching. Ideally, furniture should contain no fabric or painted surfaces. Upholstered furniture should be avoided when possible, although at times it is demanded—for example, in executive areas. Where fabrics are to be used, the hard, tightly woven materials, protected by synthetic soil retardant such as Scotchgard, are most desirable, but of course avoid the light colors such as white and light tan. Much better service is given by mottled—or mixed—and patterned colors. Woven fabrics would be better replaced with heavy plastic or rubberized upholstery material, which can simply be damp-cleaned. Avoid black or white (or very dark or very light) plastic or leather

114   Building Design for Maintainability

Fig. 52. Laminated plastic dividers and tabletops resist damage and soil and are easily cleaned. (*Formica Corporation.*)

furniture coverings, or decorative surfaces in general for that matter, since these will show every particle of dust or soil on the surface. Wiping these plastic surfaces, further, often sets up a static electric charge which actually attracts additional dust particles. Medium-color striated, or mottled, patterns require less dusting and cleaning time, just as with fabrics. Figure 52 provides an example of furniture well designed from the standpoint of maintainability.

The ideal table and desk surface is a plastic laminate such as Formica or Micarta (Fig. 53). Natural wood surfaces, although attractive, are easily damaged by cigarettes, scratching, and so on. Glass tops on desks, while providing protection, tend to cause eyestrain. Resilient desk surfaces are difficult to care for and are easily scratched or indented.

A recessed toe space (illustrated in Fig. 54) should be provided under counters, displays, "islands," cabinets, and other such fixtures. This makes it possible for an individual to approach the counter top, for example, without kicking the lower area. More important, it permits cleaning this lower area without

Furniture, Fixtures, and Fenestration   115

Fig. 53. Desirable features include minimum supports, vinyl upholstery, plastic laminate table, built-in ash tray.

marking the fixture itself. Fixtures which go straight to the floor without a toe space are often badly marked by the residue from dry mops, wet mops, and floor machines. The space should be 4 or 6 in high to permit the use of a cemented cove base without trimming it, but should be of a depth that would not collect soil and litter—2 to 4 in is acceptable.

A few specific items on furniture maintainability follow:

1. Minimize the number of legs on furniture; consider recessed bases, pedestals, and, again, wall-mounting.

2. Consider the use of collapsible tables, with chairs attached, for cafeteria areas. Carefully test samples of this type of equipment from the standpoint of safety of handling, mobility, ruggedness of construction, and collapsed size for storage.

3. Remember that furniture with glides should have stainless steel glides with a diameter of at least $1\frac{1}{4}$ in; smaller diameters cause excessive floor indentation and scoring. Use a good quality of glides to avoid the floor marking that often results from the cheaper materials.

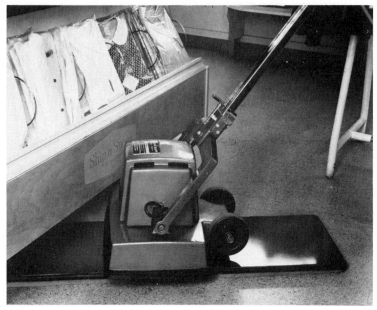

Fig. 54. Recessed bases are out of the line of vision, and the space under the display can be maintained with low-profile cleaning equipment. (*Squar-Buff Company.*)

    4. Avoid grill works, screens, and other intricate devices that are difficult to clean.

    5. Avoid unprotected wood if possible, but if it *is* used, a natural stain requires less maintenance than painting does, at least in areas where graffiti and vandalism are not a problem.

    6. Provide cabinets and drawers with deep undercuts for finger pulls, thus completely avoiding metal, plastic, or wooden handles, which invariably cause scratching and marking of the surface.

    7. See that hinges are of the concealed rather than the exposed type.

    8. On beds, avoid upholstered or fabic-covered headboards as they give constant maintenance problems, and where maintenance is not provided sufficiently often, they present a constant eyesore. Preferably, provide wooden or plastic headboards with

a smooth surface (which is also appreciated by the person who likes to sit up and read in bed).

9. For lamp bases, in tropical climates, or on ground levels where humidity or moisture may be a problem, to avoid rust use anodized aluminum or wooden bases rather than steel bases.

10. In a hospital or nursing home, avoid using suspended rods for hanging cubicle curtains, as they create a mechanical-repair and custodial problem. Curtain hangers should be recessed into the ceiling; this also avoids unsightly conditions where no curtains are being used.

11. Eliminate the space above file cabinets by furring out the wall above them, flush with the cabinet front, or by recessing the cabinets to eliminate clutter-attracting top surfaces and to expose only the front for cleaning. This is an especially important item, and worth some repetition.

FIXTURES

Water coolers, timeclocks, vending machines, and other devices are often damaged by carts or vehicles of various types. These fixtures should be protected by guardrails or wheel curbs, or by locating them in recessed areas. Recesses should not extend to the floor; there should be at least a 6-in curb.

As a matter of fact, most types of buildings use carts or other wheeled devices for transporting people, supplies, equipment, chemicals, office machines, paper, and parts. Great damage is done not only to fixtures, furniture, and equipment, but also to walls and doors by carts which have been improperly designed or selected for the purpose. Here are some guidelines:

1. The cart should not be too wide to have any difficulty in getting through the narrowest door through which it must pass.

2. Similarly, the cart should be short enough so that it can make turns easily and can fit into an elevator where necessary.

3. All corners should be rounded.

4. All corners or projections should be protected with white rubber or plastic that is securely attached.

5. Wheels should be large enough to avoid damaging floors,

the cart itself, or the item it is carrying. A 4-in wheel should be considered minimal and a 6-in wheel standard.

6. The wheels themselves should be equipped with white rubber tires.

On other types of mobile equipment which use casters, use only large, heavy-duty casters to prevent marring or damaging floors. A single wobbly caster (perhaps caused from overloading) can harmfully affect thousands of feet of flooring—*every day*.

Drinking fountains should be wall-mounted to avoid floor obstructions and other problems. The best surface material is ceramic, with second choices being fiber glass or stainless steel. Painted metal should be studiously avoided. In any case, ready access to plumbing and wiring should be available, utilizing a panel of adequate size designed for this purpose. Where an area will be frequented by children, a low-level water fountain should be provided. Where only a few adults are involved, they can use either the low fountain or paper cups, but preferably an adult's and a children's fountain should be located side by side, perhaps using a combined unit made by some manufacturers. Again, the fountain should be wall-mounted and recessed as far as possible. By all means avoid the steps which make it possible for a child to use an adult fountain, as this device creates a floor cleaning and maintenance problem and is also a safety hazard. Consider a central chilled-water system for drinking fountains to avoid the individual refrigeration equipment which would be needed for each fountain.

In wet areas, such as in food-processing plants or chemical plants, equipment and other facilities having metal bases should be elevated above floor level to minimize rusting. Equipment such as tanks, motor bases, conveyor supports, and the like, should be placed on concrete pads or masonry supports.

Where a choice exists concerning the location of fixtures which may discharge water when not operating properly (such as a laundry machine or a vending machine), especially in such buildings as apartment houses or dormitories, a basement location is much preferred to a penthouse location in order to avoid serious water damage that may affect a number of floors.

Refrigeration equipment should be located for adequate ventilation of its condenser units. Heat is removed from the item being chilled and exhausted into the surrounding air; in addition heat is developed by the mechanical operation of the motor and compressor. All this heat must be dispersed or ejected, somewhat like the requirement for electric transformer units. Heat build-up can be eliminated by ventilation, the use of an exhaust fan, or by increasing air conditioning capacity. Further, the condenser coils should be accessible for cleaning through an access door or a removable panel, as dirty condenser coils create many problems, such as high head pressure, the burning out of coil fan motors, compressor burnup, and short cycling of the compressor.

In the design of food-handling or food-processing equipment, the National Sanitation Foundation should be used as a primary source of information. In general, here are a few basic principles: Avoid sharp inside corners; use smooth butt joints; use rounded rather than sharp edges; close and weld any lips to avoid a soil reservoir; provide a means of cleaning under and behind the equipment.

Finally, consider providing slanting tops to items normally considered to be horizontal—to simplify cleaning, to avoid litter, and to avoid having people sitting or standing on them. This would be appropriate not only for certain fixtures and furniture items, but also for windowsills, lockers, radiator covers, baseboards, and so forth.

WINDOWS

Obviously, all maintenance problems with windows are completely avoided where windows have been eliminated. This fits in well with the office-landscaping technique, mentioned above. With a windowless building, we have eliminated sash replacement, hardware maintenance, problems in vandalism and security, window washing, window-covering repair and cleaning, and some other factors, depending upon the case. There are other gains, such as from sound and heat control, and other losses such as those of aesthetics and lighting, but, after all, our subject is *maintainability*. With modern air conditioning and ventilation it is perfectly feasible to construct buildings without windows, and a number of these examples exist. The objections to being "closed in" can be overcome to a great extent by color conditioning, the selection of furniture and decorations, the use of photomurals, and other such methods.

Failing the provision of a windowless building, next consider a completely sealed builing with no *operating* windows (except perhaps for cleaning purposes), as this would naturally minimize airborne dust within the structure and aid in control of its environment. This can be still further enhanced through the use of an electrostatic filtration system for the air-handling unit.

Pivoting windows—where the outside surface can be brought inside through rotation—can be an excellent investment. It permits washing the window in any weather (normally, exterior window washing cannot be done on a windy, rainy, or snowy day, or in the bright sunshine), and the washing can be done at a much lower wage rate and at convenient increments of time. Pivoting windows should be on a vertical axis, as a horizontal pivot can be hazardous from the standpoint of injury to the head or chin. Naturally, pivoting windows should be locked, and the design should be of such a type that the window is opened and closed into the same position, as this tends to extend the life of the gasket. Pivoting windows must be installed only in buildings that are on a firm foundation, as any settling will cause the windows to bind and become inoperable. For an instance of a complete refenestration with pivoting windows, see the YMCA hotel in Chicago—one of the largest hotels in the United States.

Where pivoting windows are to be installed, avoid any internal treatment that will make these windows inoperable (Fig. 55). In one building, an interior decorator designed steel valances over the windows from the ceiling down to a point that was 1 in below the top of the windows. The entire building is now equipped with expensive pivoting windows that cannot be turned! The investment in them is lost, and an expensive window-cleaning contract is now in force.

Although most people consider the overuse of glass to be expensive from a heat-loss standpoint, the consideration from

Furniture, Fixtures, and Fenestration

Fig. 55. Pivoting permits the cleaning of both sides of a window from the interior of a building. (*Aluminum Company of America.*)

a maintenance standpoint is that so much more surface is available for vandalism, and so much more time will be spent in reglazing. The amount of glass that is designed into a building should be based, then, on the factors of architectural aesthetics, heat loss, vision, cleaning, and vandalism probability.

Flat glass is considerably easier to dust and wash than is glass that is corrugated, embossed, or ribbed. The same is true, of course, for plastic surfaces. Avoid the use of exotic patterns and textures; standard patterns will avoid unsightly mixed replacements.

Large areas of transparent glass, such as floor-to-ceiling partitions at entrances, are susceptible to damage because they may be mistaken for open areas (Frank Lloyd Wright was confronted with this problem at a college he designed in Florida). This glass must either be protected, such as with an aluminum bar at waist level, or properly marked, preferably using a decal on the inside surface (the decal might consist of the company trademark or the institutional seal, or simply an artistic or geometric design).

Glass block is chosen principally for aesthetic purposes, where light transmission is desired but ventilation is not required (although windows can be installed in glass-block areas). Glass block also provides good insulation because of its enclosed air space. Economically speaking, however, it is typically not a good investment, since the same light advantages can be obtained with fixed glass (possibly tinted), using thermal glass to minimize heat transfer.

Where large windows or glass come nearly to the floor, some designs include a ledge that might contain piping, heating ducts, and the like. In public areas, such as an airline terminal, these window ledges serve as unofficial seating areas for adults and as platforms for children who then lean against the glass (Fig. 56). Either such ledges should be avoided altogether or, if the enclosed area is needed for utilities, then the cover should be sloped at a 45° angle and made of heavy-gauge material that is not easily dented. A curb at least 8 in high under the glass is recommended for all ceiling-to-floor installations for these reasons:

    1. It helps to eliminate the hazard of a person who might attempt to walk through the glass.

    2. It protects glass from floor-cleaning activities and damage by vehicles and carts.

    3. It provides an opportunity for the installation of heating and cooling equipment, as mentioned above.

Where breakage of glass panes can be a problem, such as in a public school or industrial facility where security is difficult to control, the use of plastic panes rather than glass avoids most breakage. Clear plastic can be used for windows where vision is required, translucent plastic for nonvision areas, and tinted plastic to reduce heat or glare, but it must be understood that

Fig. 56. Window ledges for heating or other purposes can be damaged when their design permits their use as a seating or standing surface.

these materials require special precautions for maintenance and installation.

For factories, warehouses, power plants, and other industrial structures, a great economy can be realized by avoiding all glass windows and roof lighting and substituting corrugated fiber glass, tinted typically blue or green. This material is very durable, will not crack, and generally looks so attractive that most organizations wash the surfaces rarely, if ever. It should be noted, however, that the material does provide problems in heat transfer, so large quantities should not be used where heat control is an important issue.

The rabbet design of the frame should be of a type that presents the best opportunity to both handle and set glass lights without damage to the rabbet or to the glass. The design must provide adequate dimensions for the glass edge encasement with sealant, and accessibility for cleaning and sealant application. The joint design should allow freedom of consideration of all types of sealants (bulk, tape, flowed-in-place, or compression gaskets) in order to conform with the compound manufacturers' recommendations, and should require a minimum number of sealing operations. Of course, the design should also incorporate a functional weep system to dispose of collected water.

Stainless steel mullions provide extensive life with little maintenance. If a vertical track is incorporated into these mullions, then window-washing gear (also used for window repairs) can ride up and down the buildings on these tracks without the need for other positioning devices.

Taking the mullion tracks further, it is possible to mount an automatic window-cleaning device on them so that the building exterior may be cleaned without human beings moving up and down on the equipment. These devices are a fairly recent development and should be investigated carefully to be sure that, for the manufacturer in question, the machine is in a practical, operating, design stage.

Where window design dictates exterior washing above the third floor (windows can be washed from the ground with a special device up to that level), a fixed track around the perimeter of the roof provides the best means of supporting a scaffolding, which hangs from a pair of cranes or davits attached to a vehicle which can be moved about on these tracks (Fig. 57). When not in use, the stage is pulled to the top, and the davits are retracted, so that all equipment is out of sight.

Considering window coverings, the most common is the Venetian blind, yet a good deal of repair (including slat, string, and tape replacement) is required, as well as continual dusting or vacuuming and periodic washing. Vertical blinds reduce the cleaning requirements but increase the mechanical repairs. Where vertical Venetian blinds are selected, heavy-duty hardware should be specified. If the appearance is acceptable, shades

Fig. 57. This automatic window-washing device is controlled by an operator who remains on the roof. (*Patent Scaffolding Company.*)

or drapery provide much less difficulty than the blinds. In one type of pivoting window, a blind is encased between two panes of glass; the theory is excellent, in that the blinds never require cleaning, but in practice this device has not yet been perfected to the point at which this investment is always a good one. Fiber glass draperies or curtains have the advantage of considerable durability, nonflammability, and—using equipment designed for the purpose—ease of washing (Fig. 58). Avoid traverse rods where possible. Use a center wand instead.

But ideally, any type of window covering should be avoided, not only to eliminate its first cost but, much more important,

Fig. 58. Fiber glass drapes launder easily and can be rehung while still wet. (*The Greenbrier Hotel, White Sulphur Springs, W.V.*)

to eliminate its repair, replacement, and continual cleaning. With the proper selection of light-controlling glass and the possible utilization of exterior shadings, avoidance becomes possible. Of course, a windowless building automatically eliminates window coverings.

Consider these window-design features and guidelines:

1. Select aluminum windows with a 1-h anodized process over wood or steel. Inside sills should be ceramic tile or cast stone. Air-conditioned buildings should use pivot-type wrench-operated windows (they also provide ventilation in case of air conditioning failure)—but consider the avoidance of window operation altogether. Windows on the south and west elevations should have external sun-control devices, such as sun screens or

louvers. All windows should be located within easy reach for periodic cleaning inside and outside or should allow easy access to both sides from the inside.

2. If windows are to be washed from the outside, avoid setting outside louvers—or any decorative masonry—too close to permit sufficient access.

3. Individual washing by a belted window-washer is the most expensive way to wash windows, but if this is required then see that hooks are built into the original construction.

4. If windows are to be flush with the exterior wall, provide a means of conducting water around the window so that it does not wash over the outside surface, leaving streaks and dirt.

5. Consider tinted, heat-reducing, glare-reducing glass (this may help to eliminate window coverings), and double-glazed windows to save on heating and air conditioning, and to improve sound control.

DOORS

Just as the best solution to the window problem may be to eliminate them completely, sometimes the best solution to a door installation situation is simply to leave out the doors. In a southern airport, for example, an expensive set of automatic doors created such traffic congestion and such a continual maintenance problem that the doors were taken down for a while; it was then found that they were not needed at all, so they were removed completely. The principal purpose of interior doors is to provide fire control, soil localization, and privacy; where these ends are not being served, doors may not be necessary, and every door location should be questioned (remember, in the "office-landscaping" concept there are no interior doors at all). Consider omitting doors to public school restrooms; privacy could be provided by a ceramic-faced wall "blind." (Be sure your building code permits this.)

Make doors large enough to permit movement of equipment and supplies without damage to the frame or to the item being moved. Watch such special cases as mechanical-equipment rooms, custodial closets, food-preparation areas, storerooms, hospital rooms. Ceiling-height doors can eliminate dust-catching transoms and trim, but the frame should be strong enough to support these heavy doors. Although a 36-in width is normally considered adequate, some state codes require other widths, such as 40 in, and these should be carefully examined. No door to a storage facility should be less than 36 in wide, and consideration should be given to the type of equipment and machinery which will be in the storage area so that a door of the proper size can be provided. Ideally, there should be double doors not less than 5 ft wide. Further, steel corner guards should be provided where frame or wall damage is to be expected, such as where beds will be moved through the door, or there will be a hand-truck or pallet traffic.

After a building has been in use for some time (usually several years but sometimes as little as several months) a decision may be reached to change the floor covering, either in selected areas or throughout the building. For example, a low-pile carpeting might be changed for a thicker carpet; or more likely a terrazzo floor which has developed cracks or a resilient floor which has become unsightly will be covered with carpet. Doors which were originally specified for the first surface will now have to be reduced in length; if they are wooden doors, a good deal of sawing and refinishing is required, and if they are metal doors it may even be necessary to replace them. Thus, where possible, doors should be installed that will permit a thicker floor covering. In many cases, a door that is too short by half an inch, for example, does not offer any problem in air exchange (but may in privacy). Another approach is the use of thresholds which can later be removed, although these create other types of difficulties as indicated elsewhere.

Exterior doors may receive much damage due to heavy traffic and weather conditions, and special consideration should be given to a number of factors (see Fig. 59):

1. Remember that where doors will be exposed to severe weather and wind conditions, a great deal of money will be spent on their maintenance unless they are protected by vestibules, recessed entrances, canopies, or overhanging roofs.

Fig. 59. Intricate exterior door framing means either much maintenance time or an unsightly appearance.

2. To avoid corrosion damage and repeated paintings, install outside entrance doors of aluminum or stainless steel. If closers are provided, they should be installed on the inside. Major entrances should provide an airlock of a double system of doors with at least a 7-ft distance between each set; each set must have two doors at least 3 ft wide. Aluminum exterior doors should be of the wide, heavy-duty type (avoid narrow-line or intermediate types). Automatic entrance doors almost completely eliminate hand smudges on push plates and glass but should be actuated by tread controls rather than photoelectric controls.

3. Remember that pulls on exterior doors should be of a design that will not create a lever action at a point of attachment to the door. Parallel horizontal bars connected by a vertical

pull prove most satisfactory and do a good job of protecting any glass in the door.

4. Wherever possible, avoid mullions, not only at entrances but at other doors as well. The presence of a mullion may eliminate the possibility of using productive equipment, such as automatic scrubbing machines for floor care, which are typically 32 in wide for corridor use in buildings. Where mullions are used, each door should be wide enough to handle such equipment.

5. Remember to design doors, including revolving doors, with consideration for floor grating.

6. Where glass doors are desired for entrances, frame them in aluminum or stainless steel to avoid damage to the door and to avoid safety hazards. Glaze them with tempered glass.

7. On outside doors, be sure to provide a slight grade away from the door to prevent both water backup and icy spots that tend to form in winter. Provide a weather drip on outside wooden doors if they are not protected from rain or snow by a canopy or wide overhang.

DOOR DESIGN

In other than residential buildings, door hardware receives very heavy use. Its replacement is expensive and time-consuming. Therefore, it is best economy to install heavy-duty hardware throughout. The finish on the hardware should be of a type that does not require either replating or hand-polishing. Probably aluminum is the best investment, as a simple cleaning is all that is required, and the material tends to blend in well with other building surfaces—although some people do not like the scratching which takes place on aluminum and prefer the dull chrome (26D finish) or stainless steel (32D finish).

Where through bolts are used to attach closers and other hardware to doors, provide spacer sleeves to prevent the collapse of the door when the bolts are tightened.

Wherever possible, the door closer should be on the hinge side of the door, thus avoiding the need for a bracket (which may require maintenance and replacement) and also eliminating a safety hazard, as many people have had their heads injured on such brackets. Floor-type closers should be avoided completely, if at all possible, since they tend to become fouled with dirt, refuse, and scrub water, and require a good deal of attention. They may also be safety hazards. Semiconcealed closers should not be used. Prevent closers from hitting walls or other surfaces when doors are opened to full swing; minimum full swing should be 90°.

Insofar as possible, hardware types should be standardized throughout the building. That way only a relatively small number of standard parts need be kept on hand for replacement. Hardware should be of a type that is easily adjusted or repaired.

If possible, the door stop should be mounted on the upper edge of the door itself, the place least apt to interfere with cleaning, and least apt to be bent or knocked off. Floor-mounted stops are easily damaged, create safety hazards, and interfere with floor cleaning, even to the extent of damaging floor machines. Wall-mounted stops are a better choice, but still provide some of the same difficulties. A hinge-type stop is a good choice where a heavy-duty hinge is used and the frame and door are both of rugged construction.

By far the most successful materials for both door push plates and door kick plates are plastic laminates, such as Formica and Micarta. Metal plates, especially those which require polishing, such as brass, should be avoided. Full-width kick plates and push plates can save much cleaning and repair of doors which are heavily used. Further benefit is obtained where the doorknob is located within the push plate, so that this area is also protected.

Wherever possible, exposed screws should be avoided; not only are they easily tampered with, but they tend to loosen and they become unsightly with soiling. Hardware should be selected that is firmly fastened to the construction and that uses a system that resists tampering and loosening through shock or vibration.

A few details concerning door design follow:

1. Remember that flush doors provide no opportunity for dust to collect in the many ridges and indentations found on

other types of doors. The simplest form of trim should also be used with the same consideration in mind.

2. On aluminum doors, avoid the narrow-line series.

3. See that side stiles on aluminum doors are $1\frac{3}{4} \times 3\frac{1}{2}$ in minimum to minimize vandalism and other damage.

4. Remember that wood doors should be $1\frac{3}{4}$ in thick and typically solid core for other than light-duty applications, such as closet doors. Solid-core doors withstand more damage and also provide better sound control.

5. Consider that although interior doors may be of wood or metal, all frames should be of metal, as wood is likely to lose its structural integrity.

6. Provide full-length glass panels in doors or adjacent to doors with a cross bar to prevent damage and accidents; failing this, provide them with an appropriate decal affixed to make the glass visible.

7. Avoid the use of louvered doors for decorative purposes only (these are typically wooden doors). Such doors require an inordinate amount of time for painting, repair, and cleaning.

8. Use rubber swinging doors in industrial or warehouse buildings where carts or fork trucks are required to push their way through the opening, or where this will take place even accidentally. Regular metal doors will not stand such treatment and will require constant maintenance. (See Fig. 60.)

9. See that doors to mechanical rooms, custodial closets, transformer vaults, telephone equipment rooms, and elevator machine rooms, are equipped with self-locking locks with free knobs on the inside, and with access by key only.

Lastly, consider two items of control. A good key-control system will result in considerable economy of operation. A metal key cabinet should be initial equipment for every building. Hardware manufacturers should be consulted concerning the master system and type of keys required.

A simple but comprehensive door numbering system is a big time saver for maintenance. Where the numbers are used for public identification, a plaque should be attached above the door, either on the casing or immediately above it. Where the numbering is for use by maintenance personnel only, it may

Furniture, Fixtures, and Fenestration   133

Fig. 60. Rubber swinging doors are designed to withstand rough treatment by fork trucks and other vehicles. (W. B. McGuire Company.)

be stamped in an unobtrusive portion of the door. For public use, use a laminated plastic plaque with rounded letters 1 in high; attach it with screws of the nonremovable type (cemented types are too easily damaged or removed and leave an unsightly scar). Where pilfering or vandalism is a problem, painting the numbers on doorframes may prove more economical. A single door number can provide a lot of information: For example, door number 6W13 identifies room number 13 on the west wing of the sixth floor.

CHAPTER SEVEN

Restrooms, Plumbing, and Piping

There are buildings without roofs, buildings without corridors, buildings without windows or doors—but a primary requirement for every building of any size is restroom facilities. As a matter of fact, it is not simply a question of need, it is a rather closely controlled legal requirement.

Restrooms and their related facilities (such as locker rooms and lounges) are "display" or "emphasis" areas; everyone sees them and uses them regularly, they judge the entire facility and its management by the condition of the restroom—and take it all quite personally. Restroom use is more or less a leisure activity, or at least leisurely, and each user becomes a self-appointed inspector with a lot of factors to claim his attention: odor, litter, graffiti, vandalism, cleanliness or appearance, provision of supplies, staining and discoloration, and his or her impression of sanitation.

But we are involved in a paradox: While the restroom is the primary area of public scrutiny (the only other areas approaching this in importance are the lobby and food-service areas), at the same time, to the building owner, the restroom is an area and expenditure that produces no income—it's just a necessary evil, and therefore of limited importance.

There is also the problem of finding people who will do a reasonable job of cleaning this type of area. The status or image problem of custodians has long been recognized as one of the big problems in motivating this type of worker; these problems are magnified when it comes to cleaning restrooms. Organizations which attempt to have a person spend full-time cleaning restrooms find that this worker is often looked down upon, even though the rate of pay may be the same and the work requires about the same skills as other custodial activities. Some contract cleaning companies capitalize on this problem and offer to clean restrooms on a contract basis so that the building owner is able to recruit his own workers for other custodial activities.

Another factor in making restroom care expensive is the fact that it requires considerable attention. Whereas offices and corridors may be cleaned only once a day, restrooms may require attention two or three times that often. In large, heavily used restrooms, such as one might find in an airline terminal or a sports arena, multiple or continuous care is needed.

All these reasons tend to underline the special importance of design for maintainability in restroom and related areas.

We should begin with layout. A number of factors are involved in deciding the location and size of restroom facilities, such as construction cost, plumbing, time lost in walking to the facilities, and so on, but from the maintenance standpoint the case is clear: Given a specific total number of fixtures, a large facility is more economical to maintain than two or more smaller ones.

Arrange the plan of a restroom—this becomes most feasible for larger installations—to provide a smooth, circular flow of traffic. The user should proceed from the commode or urinal to the lavatory, then the towel cabinet, and then the waste receptacle without having to crisscross the room. Preferably, the last step would be to use a separate exit door. Similarly, arrange shower-room and locker-room facilities to minimize floor wetting through dripping.

Individual restroom and shower-room facilities unquestionably are called for in most installations, but there are good cases for gang facilities. For example, in athletic areas gang showers are generally quite acceptable for men, and for women it is not unusual to provide a combination of gang showers and some stall showers. In industry, the circular or semicircular washbasin can be used by a number of people at one time. Each of these approaches—and there are other possibilities—results in less maintenance (as well as installation cost) than an equivalent number of individual fixtures. Full height ceramic-tile walls are recommended for gang-shower areas, but make them at least 6 ft high. Shower heads in a gang shower or any public shower should be surface-mounted with no adjustable devices; shower heads mounted on stems are too easily broken. Adjustable devices require too much maintenance. The shower heads can be mounted at varying heights for more convenient use by taller or shorter persons.

A very general problem in educational and other institutions is the maintenance of restroom facilities which are used by students, patients, or others than the staff. The problem is greatly minimized where both staff and students, for example, use a common facility, which tends to provide a regular monitoring of that facility.

To provide a most flexible arrangement, consider arranging men's and women's restrooms end-to-end with a movable partition to be changed as the proportion of men and women users changes, or the use of a double partition that puts a number of units out of commission if occupancy is very low. This would also provide for common use of facilities later on.

In certain types of public restroom facilities, such as public schools, training institutions, mental institutions, and the like, a doorless entrance to the restroom should be considered. Not only does this eliminate the maintenance of a door which would be the most heavily used in the facility, but it also tends to reduce vandalism, littering, and graffiti within the area. Of course, a blind wall should be provided, preferably faced with ceramic tile, but such blinds are necessary even with doors.

The most productive method of cleaning restrooms—all surfaces, including walls, partitions, fixtures, floors, and all the rest—is spray cleaning, where a detergent solution is sprayed onto these surfaces from a pumping device designed for this purpose, rinsed down with clear water, and then squeegeed from the floor to the drain. This activity, which saves a great deal of cleaning by hand, cannot be successfully performed in restrooms improperly designed. The basic considerations are the floor drain, the sloped floor, and wall, partition, and fixture surfaces that are not damaged by regular and copious amounts of water.

The control of odors in a restroom derives from proper design to eliminate porous surfaces in which bacteria may breed, an adequate sanitation program, and good ventilation (see Fig. 61). Exhaust fan and duct should be sized with the assumption that its efficiency will drop over the years; the covering grill should be of aluminum or stainless, as successive paintings tend to clog the openings and reduce the flow of air. The restroom

Fig. 61. The use of drip-fluid cabinets damages walls, creates an unsightly condition, and does not contribute to cleaning or elimination of odors.

door should be provided with a grill and/or the bottom of the door should be set to such a length that an air space is provided below. The fan should be on the same switch with the lighting, to ensure operation whenever the facility is in use. The number of air changes required for restrooms is typically controlled by local building codes, but they should be considered as a minimum.

To summarize points made on other subjects and add a few more, graffiti in restrooms (and other places) is difficult to eliminate but it can be lessened in these ways:

 1. Provide surfaces that are difficult to write on: ceramic tile walls with sealed grout, or epoxy-resin coating on concrete-block walls, or metal surfaces that have been given a smooth or waxed finish.
 2. Provide bright illumination.
 3. Use light-color enamel paints.
 4. Design longer toilet stalls.

RESTROOM SURFACES

Probably the best restroom floor is ceramic tile, unglazed, with a colored synthetic grout. If a portland cement grout is used, it should be properly sealed, and preferably of gray or tan. In industrial areas, or where economy dictates, concrete floors are acceptable, and again should be properly sealed. Terrazzo floors, while physically durable, are difficult to control from a staining standpoint and require excessive floor-care time. Any form of resilient floor, for anything other than private use, is an extremely poor choice: the floor is quickly damaged by water, odor control is impossible, and germs grow in profusion (Fig. 62). Carpet is acceptable for very light-use situations. The anteroom, when one is indicated, should be floored with the same material as the restroom itself to avoid a completely separate cleaning procedure for this small area. The restroom floors should be on the same level as the corridor floor, neither raised or depressed. With a proper floor-drainage system, it should not be necessary to have a high threshold, which creates numerous problems.

Fig. 62. Resilient floors, such as vinyl-asbestos, are an extremely poor choice for restrooms or other floors which may become wet.

Restrooms for active use should be equipped with a floor drain, with the floors sloped to the drain. Even if the use of the drain is not required for regular floor care (that is, where a flooding system is not used), it can still be a good investment to forestall widespread damage due to a fixture overflow, a broken pipe, or some other source of a great deal of water. In a large restroom, further consider a recessed floor receptor without curbs, in combination with a utility faucet and floor drain. (The problem of sewer gasses venting into a restroom through a trap in which the water has been allowed to evaporate is best overcome by the custodial activity of adding water to the trap periodically, or by adding a small amount of oil—such as mineral oil—to act as an evaporation seal.)

The ideal restroom wall is glazed ceramic tile. Again, as with the floors, the grout should either be synthetic plastic material or sealed portland cement, and here colored grout is still important. If the ceramic wall does not extend to the ceiling, then the upper portion should be plastered and painted with high-gloss enamel. If the restroom is constructed of concrete block, then a height of up to at least 5 ft should be covered with epoxy resin, protecting not only the block but the grout between.

Stall partitions, preferably of metal or laminated plastic, should be wall- and/or ceiling-hung, without any floor supports, as these complicate floor maintenance and require a good deal of cleaning themselves. But if the partitions are metal, they should have a finish that resists marking. Ceramic stall partitions may be

Fig. 63. The fiber glass-reinforced polyester tub and enclosure eliminates the typical problems with loose tiles and grout. (*American Standard Company.*)

used where much vandalism or graffiti is expected, but they must be floor-standing, of course, and create difficult cleaning and repair problems if they are damaged. Stalls of longer length tend to minimize the amount of graffiti or damage to the inside of doors simply because they are more difficult to reach. Although not necessarily in every restroom, do provide some restrooms with stalls of suitable width, including the door (which must open out), for wheelchair use; special provisions include a high seat, grab bars, and an accessible location.

A tub and enclosure constructed of reinforced fiber glass eliminates typical problems normally associated with ceramic construction: loose tiles, cracked tiles, broken or discolored grout. (See Fig. 63.) But care must be taken not to use abrasive cleansers on this type of material, as it scratches easily.

Be sure to specify a tissue dispenser that holds two rolls (Fig.

Fig. 64. Tissue dispensers should be of the double-roll type, so that there is always some in reserve. (*Georgia-Pacific Company.*)

64); preferably the reserve roll should remain covered until it is needed. The use of a double tissue dispenser cuts down considerably on the number of service calls and complaints resulting from running out. In women's restrooms, consideration should be given to providing a hinged bracket suitable for holding a pocketbook or package, and many women demand a construction that provides modesty strips to cover the openings between the door and the partition. In all restroom stalls, be sure to include coat hooks.

It is not enough to hang toilet-room partitions—they must be hung well. To say that the hanging should be done simply "per manufacturer's specifications" disregards the question of what the partitions are being fastened to. If the fastening is to dry wall, plaster, or concrete, a small $1\frac{1}{4}$-in shield and screw simply will not do the job; toggle bolts and more likely through bolting would be required.

WASHROOM FIXTURES

Lavatories should be of ceramic construction (not enameled steel), preferably with a back splash-panel as an integral part of the unit. The lavatories should be wall-mounted, as the presence of legs complicates sanitation. In public areas such as transportation terminals, consideration should be given to self-closing faucets. The lavatories should be large enough to prevent splashing onto the floor, and where the facilities will be used by youngsters at least one lavatory should be mounted at a lower height. Where user-acceptance is not a factor, the use of a single faucet in a lavatory, providing water of a regulated temperature for hand and face washing, cuts faucet maintenance in half while reducing faucet leakage by half, in addition to reducing first cost. To avoid floods, do not provide lavatories with stoppers. Avoid foot-pedal-operated faucets (except in certain medical areas) as they cause considerable mechanical difficulties. In public toilet rooms do not place a shelf over the lavatory as this causes a difficult cleaning problem, as well as being an accident hazard. Many people have injured their foreheads striking such a shelf while bending over to wash their faces.

The ideal urinal is wall-mounted, with integral side panels of the same ceramic material from which the urinal is constructed, and with a flooded open-throat drain connected to drain piping of large-enough size to dispose of a cigar or a paper towel. The flooded open-throat (also known as the "commode type") urinal avoids the built-in ceramic or screwed-on metal strainer, which is a constant source of odor, almost impossible to correct. (The use of rubber or plastic screens is most unsightly and odorous.) A further advantage of the commode-type urinal is that the user will much more often flush it because of the visible discoloration of the water, whereas the "dry throat" types of urinal will be much less often flushed, especially in public-use areas. The floor-standing urinal (generally not legal) should be scrupulously avoided, as floor staining becomes inevitable. (See Fig. 65.) In areas to be used by the general public, be sure to provide at least one urinal at a lower height for youngsters. Regardless of the type of urinal, a wall-mounted cigarette urn should be placed between every two fixtures to avoid their being used as waste receptacles. To avoid the unsightliness and odor caused by unflushed urinals consider either an automatic, timed flushing device for all fixtures, or the proximity device which causes flushing when the user steps away from the fixture. (Also note the lesser problem with commode-type urinals, mentioned above.) Consider eliminating metal partitions between urinals; their continual corrosion is a constant problem. Women's urinals are available and may be considered for certain installations.

Wherever possible, eliminate the use of flush tanks for urinals or water closets as they expose much mechanism and surface for damage, soiling, and vandalism. Use flush valves instead, with flush handle at least 36 to 42 in above the floor, where they are easy to reach with the hands but difficult to reach with the feet—thereby avoiding a good deal of soiling and damage.

Water closets should have the following features:
    1. Wall-mounting, to avoid cleaning problems on the floor
    2. Elongated bowl
    3. No tank

144   Building Design for Maintainability

Fig. 65. Floor-standing urinals, drain strainers, and proximity to metal walls all create maintenance headaches.

    4. A flush valve 30 in above the finished floor to avoid damage to the valve from being kicked with the foot

    5. No floor valves that are foot-operated; no seat-back valves

In the selection of toilet seats, a number of factors can reduce or eliminate maintenance problems:

    1. If no seat cover is provided, a daily cleaning of both the top and bottom surfaces is eliminated, as are problems with the hinge and the bumpers.

    2. An open-front elongated seat not only reduces sanitation time, but also generally provides a better appearance. Split seats should be of rugged construction to avoid damage in the hinge area. Split seats are recommended for both women's and men's restrooms because of general sanitation, the possibility of conver-

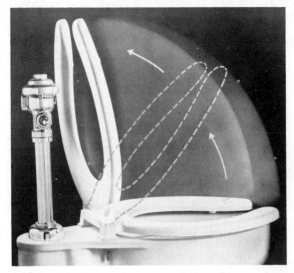

Fig. 66. The self-raising toilet seat helps solve one of the most common problems in restroom maintenance. (*Beneke Corporation.*)

sion, and the need to stock only one type of seat for replacement purposes.

3. The self-raising seat provides considerably better sanitation conditions and again reduces cleaning time. (See Fig. 66.) The newer models cause the seat to raise slowly, and have concealed springs. Exposed springs should not be used because of the cleaning problem.

4. Solid plastic seats provide great durability and almost indefinite life, contrasted with coated wood or pressed fiber seats which may tend to chip or split.

5. Black bumpers should be avoided, and either white rubber or plastic bumpers specified.

6. In certain institutional situations, a water closet may be provided without a separate seat; the fixture itself has its upper surface finished in the shape of a seat.

In an industrial area, if a circular or semicircular hand-washing fountain is provided (such as the "Bradley basin"), avoid the terrazzo basin, as it is easily damaged by chemicals or other

Fig. 67. A good first choice for a circular washbasin would be reinforced polyester-molded construction; a second choice would be stainless steel; a poor third choice would be terrazzo. (*Bradley Corporation.*)

means. The fiber glass or stainless steel basin are better choices, in that order. (See Fig. 67.)

If no custodial closet or utility sink is available, provide a wall-mounted bibb-type faucet (so that a hose may be screwed onto it), with a supporting bracket so that a bucket may be hung on the faucet. This should be located over a recessed area of floor, or a curbed area, with a drain in the deepest part of the recess. (Do not forget the siphon breaker.)

For bathtubs and shower stalls, specify fixtures that have built-in antislip features. The later provision of adhesive strips or spots requires regular repair and replacement and makes fixture cleaning more difficult. The alternate system of providing nonslip bathmats leads to theft, a good deal of handling time on the part of housekeeping personnel, and still another item to be cleaned.

Avoid shower curtains to eliminate a continual cleaning, repair, and replacement problem. Shower partitions and doors should preferably be of aluminum and fiber glass. If glass is to be used, it should be of the tempered type.

RESTROOM EQUIPMENT

The modern approach to restroom equipment is to provide recessed, multipurpose units of large capacity and made of durable materials. The recessed construction eliminates surfaces to be cleaned and makes vandalism more difficult. The largest storage capacity for towels and soaps minimizes replacement time and service calls.

Some varieties of restroom receptacles, such as stainless steel units recessed into the wall, have a locked lower portion. Periodically these are damaged—often beyond repair—by someone who has inadvertently dropped an item of value into the receptacle. One would not think this would happen regularly, but people drop in anything from false teeth to watches; their efforts to retrieve these items result in bending the door to the receptacle. To avoid this problem, the original installation should include a sign cemented or bolted to the fixture, perhaps of embossed laminated plastic, indicating the number to call for the key.

All public women's restrooms should be provided with a wall-mounted recessed sanitary napkin dispenser as well as a wall-mounted receptor in each commode stall area. In fact, wall-mounted dispensers and waste receptacles in general are important in avoiding rust, floor-cleaning difficulties, and the extra attention floor-standing containers require (see Fig. 68).

Avoid mounting mirrors over lavatories as this may lead to piping stoppages due to loose hair. Further, avoid mounting mirrors with rosettes as they are hard to clean around. A good solution to the mirror problem is to have them as an integral part of a combination recessed wall unit with mirror on top, towel dispenser beneath, and waste receptacle at the bottom. In women's restrooms, consider the desirability of at least one full-length mirror. (How does this affect maintenance? More vandalism, litter, and graffiti occur where persons feel they have not been given adequate facilities.)

Fig. 68. Where inadequate provision is made for waste receptacles, perpetual problems of litter, fire hazard, vermin, and difficulty of floor cleaning result.

There are literally dozens of varieties of hand-cleaning materials, including such basic items as bar soap, powdered soap, lotion-type soap, liquid soap, soap leaves, soap papers, and so on. The two systems providing the best results with a minimum of maintenance and attention are liquid soap, which is dispensed in lather or foam using a special valve for this purpose, and lotion-type soap. In any case, avoid glass soap containers; use plastic or metal. Avoid foot-operated soap valves; the only justification for such equipment is in surgical preparation suites and then a substitute system can be used. It is not economically feasible to mount a soap dispenser between two lavatories as an economy measure as the soap (liquid or powder) that falls to the floor will cause a considerable custodial and safety hazard (Fig. 69). A dispenser should be mounted over or within each lavatory.

If bar soap is to be used in the lavatory, a drained indentation should be part of the lavatory fixture itself, eliminating the need for a separate soap dish. In a tub or shower, provide a combination soap dish and grab bar of the same ceramic material forming the tile wall; it should be the self-draining type, both for the

Restrooms, Plumbing, and Piping 149

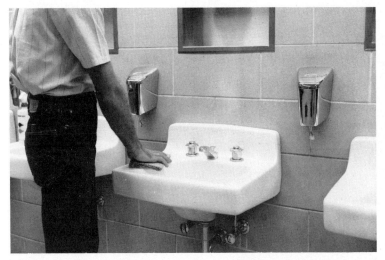

Fig. 69. Where soap dispensers are mounted between lavatories, material falling on the floor will cause staining of the grout and a regular sanitation problem.

ease of cleaning and to avoid the dissolving of synthetic detergent bars.

A central distribution system for liquid soap can be a good investment for large high-rise office buildings, where numbers of restrooms are involved. Where each restroom is typically small, having a half-dozen lavatories or less, a system might be considered for the entire building or for several floors of the building. Where each restroom is rather large (a dozen lavatories or more), a central system for that individual restroom may well suffice and be most economical. Many central liquid-soap systems have been in use for years without causing problems. (The most important operating factors are keeping the pipes full and never introducing into the system liquids of other than the proper dilution.) Naturally, only an economic analysis for any given facility can indicate the desirability of a central system.

Although cloth toweling is often preferred in executive areas, it is not the best application for general use. Paper toweling

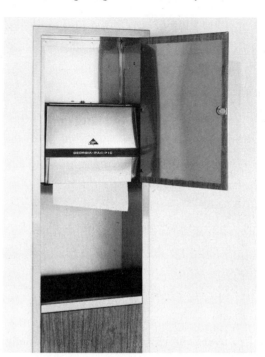

Fig. 70. Most desirable for restrooms are combination recessed fixtures providing toweling, waste receptacle, and mirror. (*Georgia-Pacific Company.*)

supplied in rolls, dispensed in controlled amounts, is an economical system. Theoretically, electric hand driers have a great deal to offer, since there are no towels to replace and there is no litter to remove, but in practice the vast majority of organizations that do install electric hand driers later come back and put up paper-towel dispensers alongside. This is simply because electric driers are too slow for most people and also simply cannot be used for drying the face or the back of the neck. Whatever type of system is used, whether cloth, paper, or electric, at least two units should be installed to eliminate complaints and allow for the time required to refill dispensers (or repair the electric drier) on an emergency basis. Similarly, the waste receptacle should have at least one companion unit. The doubling-up of

Restrooms, Plumbing, and Piping 151

dispensers and waste containers is a good investment for anything other than private restrooms used by just two or three individuals. The towel dispensers and waste receptacles as well, which may be a combined unit, should be wall-mounted and recessed as far as possible (Fig. 70). Avoid painted steel cabinets; stainless steel, while not perfect, is a good material for this purpose.

Lockers should be mounted on a solid base, such as concrete, with a ceramic cove; this provides easy cleaning, and eliminates a space where litter and soil might collect (Fig. 71). The base

Fig. 71. The solid base eliminates a litter trap; the integral seating eliminates the obstruction of supports; but lockers should have sloping tops to avoid litter collection and the floor should have been ceramic tile.

on which the lockers rest preferably should be of solid material, such as poured light-weight concrete. If it is hollow, there is a possibility that water and soil will collect, creating odors from the growth of bacteria that will be impossible to reach and remove. Further, the base should be recessed as a toe space to a depth of about 2 in and a height of a 4-in cove to avoid mop marks and rusting of the lockers themselves, which would happen if the front of the lockers are flush with the base. The tops of the lockers should be slanted to avoid collecting soil and having numerous things piled on top. Preferably, locker-room seats should be cantilevered from the lockers themselves, so that no floor legs are required.

PLUMBING

A discussion of restrooms naturally leads one to a consideration of plumbing, since the majority of plumbing requirements in a building are for this purpose.

In addition to the general drawings showing the plumbing system, special care should be taken to show the location and sizing of all underground pipelines. At junctions or turns in the line, of course, manholes should be provided. Careful marking of underground systems is especially important where plastic piping is used, because electronic location gear will not function.

Plumbing and piping systems that are properly identified avoid wasting the time required to search out and test directions of flow, contents, and hazardous situations. This identification should be affixed to or painted on next to each valve, each branch or take-off, each point a pipe leaves or enters a wall, each major change in direction of a pipe, each expansion joint, and each anchor. A brass, stamped tag should be used to identify valves.

In addition to other identification, the color-coding system has been found to save a good deal of time and to eliminate some accidents. Color coding can be used either by completely painting pipes or pipe coverings or by using colored bandings at frequent intervals. The following schedule (developed by the American National Standards Institute) is in regular use:

Fire protection	Red
Steam supply	Orange
Condensate return	Aluminum
Hot water	Dark yellow (gold)
Cold water	Dark green
Chilled water	Light green
Condensing water	Light grey
Gas	Light yellow
Air	Dark brown
Drains	Natural with walls
Electrical	Natural with walls
Vacuum	Beige

The value of standardizing fixtures should be of considerable economic weight in purchase considerations. Certainly all fixtures within the same building should be of the same type and from the same manufacturer, and the same thing is also true in facilities having a number of buildings. For example, where all restroom fixtures, valves, and fittings are standardized (even if they have been purchased at five- or ten-year intervals) a stock of replacement parts and whole units considerably reduces the time and cost of repairs or replacement; but where such items have been purchased from several manufacturers, it may not be feasible to stock such parts, and high costs will result not only from the increase of maintenance expense, but also the down time imposed.

In determining the space necessary to properly maintain piping and plumbing equipment, we must assume that every fixture, valve, or fitting will eventually have to be removed or worked on. Where inadequate space is provided to swing a wrench, and to perform the proper body motions, the worker will require excessive time due to fatigue, cramping, and giving attention to avoiding skinned knuckles or other accidents. (See Fig. 72.) Specifically, for fixtures with screws, be sure that the item to be removed has enough turning room; for bolted flanged fixtures, be sure that every bolt can be turned with a wrench. To the plumber, plumbing chases seem invariably cramped; 3 ft should be considered as a desirable minimum width.

In piping layouts, be careful not to block openings through

154   Building Design for Maintainability

Fig. 72. This congested maze guarantees that any maintenance activity will be prolonged and expensive.

which equipment must be moved. Even though it may be intended for the equipment to be rerouted elsewhere, the piping would probably become damaged by attempts to force equipment through. Keep piping, valves, controls, operating devices, hangers, and other permanent fixtures out of aisles and access space where they may be damaged by vehicular and other traffic; but if such location is unavoidable, provide guardrails and bumpers, along with cautionary painting.

A maintenance worker's productivity will be reduced with exposure to such hazardous conditions as hot piping, dripping liquids, and the like. Provide insulation for steam and condensate piping, at least the amount that will be exposed to human contact; secondary benefits include reduced compensation costs and utilities cost.

Floor drains should be provided—and plumbing roughed-in to assure location at the low point of the floor—wherever fluids might accumulate on a regular basis. Of course, drains should be adequately sized to eliminate clogging. Examples would include the following:
1. Food-processing areas
2. Steam-cleaning areas
3. Restrooms of considerable size and/or where the spray-cleaning technique is anticipated
4. Wet-processing areas
5. Safety showers
6. Parking lots
7. Sprinkler test drains, if inside the building
8. Areas with large floor gratings to be serviced
9. Pop-off valve discharge areas, where the discharge cannot be piped to the exterior

Waste drains for restrooms—urinals, water closets, showers—should be adequately sized to prevent clogging, as overflows can cause very expensive problems.

Major mechanical and air-conditioning-equipment rooms should be provided with floor drains to accommodate leaks, broken pipes, discharges, and provide for on-the-spot cleaning of parts and piping. Where possible, discharges should be piped to the outside of the building rather than to the floor drain. The best investment for water piping is red brass; galvanized piping will become extremely expensive as it corrodes and becomes clogged with scale. For piping in general, plugged tees rather than elbows will provide opportunities for drainage or cleanout at very little additional cost.

The typical piping system does not have enough valves, so that a large portion of the system needs to be shut down when maintenance is required. True, valves are expensive, but they more than pay for themselves on a single shutdown. For example, each runout from a riser should be accessibly valved, and zone valves should also be accessible. Water lines should be valved at the point of connection to the water main, and also at the entry to each building, both inside and outside.

Wherever bibb-type faucets are used (with screw threads so

that flexible hoses may be attached), siphon breakers must be provided in order to avoid the suction of impure water into the plumbing system. Typical areas for this would include the kitchen, custodial closet, boiler room, laundry, production areas, and the like.

Pipe unions should be provided at each valve for quick removal without the necessity of repiping. No valve should have its handle pointing downward.

Valves should be standardized as far as possible, and spare parts kept on hand, as well as a few whole spare valves. The same is true for other critical parts, such as unions and strainers.

Where overhead valves or other devices require frequent resetting, the installation of chains or extension handles can make this resetting possible from floor level, thus avoiding the need to obtain ladders or lifting devices.

The location of toilet flush valves 24 in above the top of the closet can avoid a good deal of damage from the common practice of using one's foot to operate the valve. (This practice stems from the user seeing, or believing, that the valve is not kept clean.)

To provide proper maintenance to lavatory faucets, each hot- and cold-water riser should be individually valved near the bottom of the riser.

In cold-weather areas, outside faucets and bibbs should be of the freezeproof variety, or encased in a box. It is a great waste of time to have to wrap standard faucets to prevent freezing, and a broken faucet requires shutdown of the affected water system, the time and expense of replacement, and even then the released water may have caused a good deal of damage.

Hydrants should be provided that permit withdrawal of internal working parts without disturbing the barrel or casing valve when shut, and should be reasonably tight when the upper portion of the barrel is broken off.

The copious amounts of water given off by sprinkler heads during a fire can be of the greatest benefit in protecting life and saving the building, but when there is no fire a sprinkler head that goes off creates a number of serious problems from water damage. A device is available, consisting of a form of wedge on the end of a long handle, which makes it possible for a person to shut off the flow of water while standing at floor level. This could be especially important in multistory buildings where run-down on ceilings and floors below might damage valuable surfaces and materials; where floors are resilient, carpet, or wood, and thus very susceptible to water damage; or in hazardous areas where the shutoff time for the portion of the sprinkler system controlling the broken head must be minimized. Consideration should be given to locating sprinkler stoppers in initial construction or renovation so that the location can easily be remembered through association with fire or water, such as next to every water fountain or every fire-alarm box, or both. The more the sprinkler heads are subject to damage, such as in low-ceiling areas where materials-handling takes place, the more significant does this protection become.

Where leakage is a problem from pumps and valves or other equipment, or where leakage could become a serious problem in a system containing corrosive or hazardous fluids, provide corrosion-resistant drip pans at sensitive locations. For larger installations these can be connected with the drainage system.

Laundry chutes and waste chutes should be equipped with a floor drain at the bottom and a cleaning-spray head at the top.

In hard-water areas, consideration should be given to the installation of water-softening equipment. This provision would considerably reduce the scaling formed in piping, drinking fountains, on the surfaces of lavatories and other restroom fixtures; it would also help to provide water for preparing detergent and disinfectant solutions which would perform more effectively and, at the same time, use less of these chemicals.

Some details concerning piping:

    1. Wherever possible, install all pipe work (which includes steam, condensate lines, and traps) in a manner which will permit the condensate to drain by gravity from the steam side to the return lines, thus keeping the steam lines free from condensate when the steam pressure is off.

    2. Where a pipe passes through masonry construction, fit it through a metal sleeve so that neither the pipe nor the wall

is damaged from motion of the pipe caused by vibration or expansion and contraction. For uninsulated pipes, the sleeve diameter should be at least 1 in larger than the pipe, while for insulated pipes it should be one pipe size larger than the outside diameter of the insulation. Where sleeves pierce an outside wall, they should be made of galvanized steel pipe with a water-stop flange to prevent water entering the building from that source.

3. Remember that where 2-in or larger pipes are supported by rollers, a metal protection saddle should be in contact with the roller to keep the insulation from tearing. Of course, the saddle itself should be packed with insulation.

4. Remember, too, that antifriction piping supports (such as rollers or graphite slides) on long runs reduce the stress on supports and thereby reduce the damage to the support anchors, whether they be in the wall, ceiling, or floor. The antifriction supports also reduce damage to pipe joints, valves, and other fittings.

5. Provide stub-tees which are plugged or capped off in piping for future use; if the future expansion would be contemplated without the disruption of the regular piping use, the location should also include a valve.

6. Where pipes, valves, or other similar equipment is subjected to temperatures above 190°, specify a heat-resisting black enamel or aluminum paint to provide suitable and durable protection to metal surfaces.

CHAPTER EIGHT

Trades Maintenance

The division of subject matter concerning design for maintainability into chapters is difficult because a precise separation is not possible; too many subjects are interrelated. Although the subject of this chapter is trades maintenance, plumbing was covered in the restroom section because of their close relationship; custodial work is mentioned throughout, and so forth. It is convenient in this chapter, then, to consider electrical (and lighting), mechanical, heating, ventilation, and air conditioning. First, however, we had best discuss related structural items.

STRUCTURAL

For a facility which will require a considerable quantity of piping, wiring, and conduit, consideration should be given to providing a means of ready access in areas designed for this purpose, rather than concealing these utilities and requiring their rather difficult exposure when the need arises. An intermediate method is the use of access panels or plates, such as on modular centers in floors, or the use of movable pedestal floors or false partitions to permit access. (See Fig. 73.) Ideal utility exposure would be overhead in a utility "attic," from below in a "basement,"

160   Building Design for Maintainability

Fig. 73. The access panel is desirable, but it is too small and the chase too congested.

or perhaps a vertical space running lengthwise through the center of a building, which could also proceed vertically through a number of floors. The vertical utility space would serve areas on either side. The situation is analogous to having mechanical spaces located every several floors in a high-rise building, with each space serving, for example, several floors below and above. Utilities may also be clustered in corridors, either being covered with a removable ceiling, or simply being "painted out" with flat paint above a suspended lighting system. Also consider utility tunnels joining smaller buildings, or portions of a large building. Utility tunnels, utility chases, and the like should be secured from unauthorized persons: Vandals, arsonists, and others can cause havoc in these areas.

Especially in industrial facilities, service galleries can provide easy, safe access to overhead facilities for inspection, lubrication, repair, and cleaning. The gallery, or catwalk, would typically have a grating for a walking surface, metal curbs to prevent objects falling off the catwalk, and safety railing. The gallery should be wide enough and strong enough to accommodate equipment for welding, steam cleaning, vacuuming, and relamping.

Where permanent facilities require frequent climbing, install a permanent ladder as a part of the equipment; also involved may be a catwalk, guardrail, or work platform. Where regular lifting is required, install a beam or hook overhead for the use of a hoist or chain fall. Permanent ladders, catwalks, and work platforms should be located where they will not become a security problem; they may be a temptation to unauthorized persons, vandals, and so on.

In any structure over 120 ft long, provide expansion joints—for floors, walls, ceiling, piping—to avoid structural and surface damage.

I beams and H beams are much easier to keep painted and cleaned than are trussed structures. Minimize the number of exposed structural columns.

Where exposed structural steel is used for columns, such as I beams, H beams, or channels, the space between the flanges near the floor typically becomes filled with litter and soil, entangles mops, and the like. This situation can be avoided by filling in these flanges on a bevel, typically at a 45° angle, with a light-weight concrete or other material. For best results, the concrete should be sealed to provide a smooth surface.

Where conveyor systems are enclosed, provide plenty of hinged or removable access panels or cleanout doors. Typically, it would not be amiss to make one whole side of the conveyor removable, except for the supports. This facilitates inspection, maintenance, retrieving of lost items, pest control, and cleaning.

ELECTRICAL

The mistake made most often in designing electric systems is the failure to provide adequate capacity. Take into consideration not only those appliances and devices which require electric current today, but also consider the certainty of the invention of additional devices, and the proliferation of the use of today's devices for tomorrow's use.

Provide empty conduit from recessed electric panels to junction boxes to handle future requirements for electrical outlets and fixtures, various types of powered equipment, telephone and communication wiring, data processing, public address and music systems, integrated clock systems, alarms, teletype, new inventions, and so on. Failure to make such provision leads to exposed wiring which is at once unsightly, unsafe, and requires a good deal of maintenance attention.

In most buildings, areas that were originally assigned for storage or other nonadministrative purposes are later reassigned as offices. If there are such conditions as inadequate wiring, exposed molding, and the like, they create a number of problems. Consider the areas of most likely reassignment so that they may be properly wired in advance.

Where the circuitry is likely to change, such as in flexible industrial situations, research laboratories, or pilot plants, provide rack, tray, or channel construction for the support of cables and wires; this permits faster changeover at lower cost and also exposes the wiring to immediate maintenance without pulling through conduits.

In industrial plants, certain types of laboratories, pilot plants, and other areas where electrical use is heavy and the relocation of equipment is a common practice, the use of electric bus bars (or bus ducts) can avoid a great deal of rewiring, patching, hole drilling, and other installation and maintenance problems. As this is an expensive first-cost system, the economics must be carefully evaluated. Where a great deal of airborne soil is present, such as lint or paper dust, and especially where this is combined with a high humidity, it may not be possible to use bus bars.

In office areas, 120-V outlets should be spaced not more than 6 ft apart. Good practice indicates the desirability of an outlet on each wall of a private office, thus providing for the greatest flexibility in the arrangement of the furniture and in the use of powered office equipment. In lobby areas, an electric receptacle should be located on each column, and at intervals of no more than 50 ft around the periphery. There should be an electric receptacle in every elevator cab, in the elevator lobby, and on

Fig. 74. Power- and telephone-change costs are reduced through the "plug-in floor" concept. (*H. H. Robertson Company.*)

every stair landing. Each restroom and locker room should be equipped with these outlets, as well as all service, storage, and maintenance areas. The corridor should have outlets at a maximum of every 50 ft. Assume that maintenance equipment will be fitted with 1-hp motors or larger. In addition, 208- or 240-V outlets should be provided where the use of heavy equipment is anticipated. See Fig. 74.

In restrooms, provide suitable electric outlets not only for electric shavers but also for the operation of a scrubbing machine and wet vacuum. One electrical outlet should be provided high on the wall in a location suitable for the installation of an electric deodorizer. Provisions should be made for electric hand driers if they are to be used. Illumination in the restrooms should be at a high level, such as 50 fc, to reduce vandalism, graffiti, and litter. Be sure that the lighting is installed in such a way that the inside of the toilet partitions are illuminated as well as the outside; one way to do this is to mount double-tube strip fluorescent fixtures over the row of stall doors, thus lighting both sides of the doors.

Avoid canopy switches, pull-chain-operated receptacles, and power outlets in light fixtures. Switch panels, where accessible to occupants of a dormitory, nursing home, or similar facility, should be equipped with provisions for locking. Ideally, panel covers should be of the door-within-door type, with the inner door exposing the circuit breakers only, and the outer door covering the wiring; no screws should be exposed with which to remove these doors (see The National Electric Code).

Facilities for battery charging might be necessary in a number of locations for such items as automatic scrubbing machines, fork trucks, personnel carriers, trucks, and so forth.

Avoid floor-mounted receptacles; they are easily damaged by detergents and waxes, difficult to clean around, and are the cause of accidents.

Avoid running water and drain lines through electrical equipment areas (actually, this is prohibited by code) since water from even a tiny break can cause extensive damage and create unusually high safety hazards.

A number of factors concerning the location of electrical equipment are of importance:

    1. Be sure that all electrical equipment is readily accessible, so that it may be both cleaned and repaired easily and quickly. Leave plenty of room available for easy, safe work by maintenance personnel.

    2. Do not place electrical switchgear and controls in locations where they may be damaged by materials being carried about, or by vehicular traffic, or by other means. For larger installations, a separate area should be set aside, which is properly enclosed (such as in a chain-link fence), and this enclosure itself properly protected by curbs or bumpers. These enclosures also provide a minimization of electrical safety hazard. Any floor that may become wet above an electrical control room should be waterproofed.

    3. Mount electric junction boxes and other gear facing away from conditions which create chemical splashing, spraying of powder or dust, flying chips, and the like.

    4. Locate transformers and switchgear which are installed on the interior of buildings in an area that is well ventilated,

in order to prevent overheating and subsequent burnout. It may be necessary to include the area within the ventilation or air-conditioning-system area, to leave one or more openings in an outside wall, or to provide an exhaust fan actuated by a thermostat.

5. Where electricians or other maintenance personnel are required to work around exposed wiring, unprotected knife switches, or other dangerous conditions, both their state of mind and their productivity will suffer. Be sure to provide adequate insulation and protection, even in areas that are fenced in and locked. The more congested the area, the greater the protection that will be required.

6. Locate panel boards, electrical gear, and electrically operated instruments apart from areas that are subject to shock or vibration, thus considerably reducing servicing and maintenance requirements.

7. Remember that indoor electric vaults should include a drain and a sloped floor for removal of oil and water and should be provided with positive cross-ventilation. Vaults should be accessible from the exterior of the building, with a 42-in minimum-width opening.

Just as mechanical equipment, duct work, and piping should be identified to reduce maintenance time, so should electrical equipment be identified. Here, another important aspect is safety. Use laminated plastic signs, securely attached with rust-proof screws.

LIGHTING

A primary concern in lighting design is accessibility for relamping, cleaning, and maintenance (Fig. 75). In the location of light fixtures, consideration should be given to the means of getting to them for either lamp replacement or general repairs. The location of fixtures over hazardous types of chemical or mechanical-processing equipment should be avoided, and the lighting should be on the circumference of such equipment but properly directed. In high-ceiling areas, relamping can be so difficult and expensive that it just does not get done unless spe-

Fig. 75. This lighting fixture provides both accessibility and a solid cover that is easily cleaned. (*General Electric Company.*)

cial provision is made for it. In some cases, retractable lamps, or lamps that can be lowered, solve this problem.

The location of some stairwell lights—for example, 15 ft above the stairs themselves—creates such a difficult, time-consuming, and hazardous job of lamp replacement that the job is either delayed or avoided. This, in turn, causes a chain reaction of problems in vandalism, wall damage, and safety. Stairwell lights should be located either over the landing to permit the safe use of ladders, or preferably on the side walls of the landing so that they may be reached by hand. (Also see the Elevators and Stairs sections in Chap. 4.)

The use of long-life lamps for general purposes is a questionable economy but one that has to be investigated for each situation. If such economy rests on guarantees of lamp life, it is quite possible that the required record-keeping costs will counterbalance any savings. On the other hand, rough-service lamps are generally a very good investment in problem areas, such as where building vibration or shock (either mechanical or thermal) is anticipated; or where lamp replacement is difficult because of inaccessibility, the presence of operating machinery, chemical exposure, or the like.

If possible, lights on each floor should be controlled from a central lighting panel; this will eliminate individual wall switches, and the repairs, smudging, and cleaning that must go with them. Use separate switches for this purpose, so that circuit breakers are not used for routine switching.

If lighting is controlled from a central location, provide controls so that it is possible to illuminate only every other row of lights. There are a number of opportunities when this would be ample for many hours at a time, such as security lighting.

Try to arrange the lighting system to use a standard voltage. For example, a building having both 120-V and 277-V lighting may suffer time loss and bulb damage when workers attempt to put a lower-capacity bulb in a higher-capacity fixture. This will cause immediate failure or even a bulb explosion. (The opposite procedure causes no damage, but simply wastes time.) The use of 130-V bulbs for incandescent lighting rather than 120-V bulbs provides longer life, even though the illumination will be slightly reduced.

Certain areas are typically dimly lit, such as restaurants, lounges, theaters, and film-processing laboratories. These areas should have either a secondary system of bright lights or a system of intensifying the dim lights so that proper, safe maintenance can be provided.

Avoid the location of accent lights high up on columns or in other inaccessible areas, since experience has shown that they may actually be abandoned after a time because their maintenance and relamping is so difficult.

Where daily flag-raising ceremonies are time-consuming or difficult, flag-pole lighting controlled by a solar switch can help. But check state law first to see that it permits flying the flag by night under artificial lights.

Consider exterior floodlighting to minimize vandalism.

For installations having a considerable quantity of fluorescent tubes of any length, initial equipment for the central maintenance facility should include a tube crusher. This machine safely solves the problem of tube handling, which is often a hazardous activity. The volume of discarded tubes is reduced to a minimum, which expedites disposal.

A number of design features concern the lighting fixtures themselves:

1. Lighting units which illuminate upward as well as downward, and provide normal-level lighting from ceiling reflection, do not collect dirt as quickly as closed-type fixtures.

2. For lighting in a typically cold environment, such as a refrigerated food-processing area, low-temperature lamps should be specified, as they provide peak output down to 0°F. Similarly, low temperature ballasts should be specified. To provide special protection for the lamps, such as against cold or strong winds, specify jacketed lamps.

3. In general, suspended fixtures should be avoided as they provide a top fixture surface which collects soil and also require maintenance of the hanging device itself. (On the other hand, fixtures directly mounted to a ceiling which is made of cellulose fiberboard may become overheated, in which case it would be necessary to drop the fixture on short stems to provide some air circulation.)

4. Light globes, either for outside or for inside use, should be of high-impact plastic; avoid glass.

5. The accumulation of dust and soil on lighting fixtures and lamps can actually cut in half the amount of illumination normally coming from the fixture. In industrial situations where a great deal of dust and dirt is anticipated, ventilated fixtures should be considered; these create a "chimney effect" by a series of openings located on top of the reflector, thus permitting the heated air around the fixture to carry soil past the lamp rather than depositing it on the lamp. Although not entirely eliminating the soiling problem, the ventilated fixture does alleviate it. (Of course, a completely enclosed fixture would be preferable, although these are usually too expensive for industrial use.) Whenever fixtures will need frequent cleaning, such as when they are located in dusty or dirty areas, specify a type that can be easily dismantled and cleaned.

6. Fluorescent fixtures, whether suspended, surface-mounted, or recessed, are often covered with egg-crate diffusors; these items are difficult to clean and easily damaged. A solid lens of glass or (preferably) plastic has the advantage of easier and quicker cleanability. Where conditions permit, the elimina-

Fig. 76. Exposed fluorescent tubes simplify replacement and eliminate the cost of maintenance of covers or diffusors.

tion of diffusors or lenses altogether also eliminates maintenance and replacement (Fig. 76). Where egg-crate louvers are to be used to a large extent, the central maintenance work facility should be provided with either a solution tank for the cleaning of these louvers or a portable louver-cleaning machine.

7. The inner surface of which the reflector is constructed will have a great deal to do with its cleanability and soil retention, and thus the illumination output. Plastic surfaces vary widely concerning their stability and discoloration, although this is generally overcome with a regular cleaning program. Enamel surfaces are easy to clean. In many cases the best surface would be polished aluminum, as painting is not involved and washing is not difficult. The question of aesthetics would have much to do with the decision concerning fixture surfaces. Porcelain enamel is more durable than synthetic enamel but also more expensive.

MECHANICAL

Three-dimensional scale models, while rather expensive, can save a great deal of cost and avoid a number of errors in the installation of mechanical and process equipment. Clear plastic sheets can be used to indicate floor surfaces, and these can be marked with a grid to indicate 1-ft squares. Color-coded dowel sticks or miniature tubing can be used to indicate piping. Very rarely indeed is the use of such a model, for a complex installation, considered to be anything other than a good investment; and of course the model should be retained for possible use later on in renovation or relocation.

Sometimes equipment is installed in such a way that access panels, inspection plates, couplings, and other devices are difficult to get to, and sometimes almost impossible to use. There have been numbers of cases where outside stem and yoke (OS and Y) valves have been installed, which, it was then found, simply could not be opened because there was no place for the stem to go! Drawings showing the installation of mechanical equipment should show all devices in open position in dotted lines, just as in floor plans the swing of the door is so indicated. (The use of scale models, as above, can quickly point out this problem.)

Equipment spaces which are too crowded will create a considerable increase in maintenance and repair time and also lead to more accidents. Access to heavy machinery with hoisting equipment should be considered in the design stage.

To go further, consider the advantages of building an overhead crane into the utilities and power-plant building. Normally such structures do not have interior columns, and the inclusion of a crane, if the building is not too wide, might be a relatively simple matter. The economics of both a powered crane and a manual crane (which consists of little more than an I beam on wheels) should be considered. A crane would not only simplify the original installation of heavy and bulky equipment, but also its repair and eventual replacement. In some cases, the use of a crane (either built in or rented) is the only means of removing heavy equipment when it is hemmed in by other equipment. If a crane is to be used, attention must be given to any piping or duct work that goes through the roof of the building; either this should be rerouted, or a simple means provided for its temporary removal. Consideration could also be given to a roof-mounted crane or hoist, with an opening through the roof to service equipment beneath.

Mechanical (and electrical) equipment rooms should be separated spaces which can be secured from the occupants of the building. Of course, they should not be used as passageways, storage areas, or for any other purpose. Therefore, it is necessary that departmental equipment and controls (such as air compressors and distribution panels) be located in areas accessible to the occupants of the building so that they will not have to enter the major mechanical areas.

Machine maintenance can become time-consuming and even hazardous without sufficient illumination.

Installation should not be considered complete unless a proper file or card is set up on each item of equipment showing manufacturer's address, location of nearest repair facility, maintenance record, and spare parts kept on hand.

Identify equipment with laminated plastic signs (painted or stenciled signs tend to get painted over or rubbed off). The signs should describe the name of the equipment, its rating, certain maintenance instructions, cautions, and other important data.

A central panel board will save many man-hours in walking about gathering information on readings for pressure, humidity, temperature, utilities consumption, and the like, as well as provide an early alarm for problem situations, and better observation of operating conditions.

In selecting operating equipment and its components, minimize the need for special tools, gaskets, lubricants, and other requirements that complicate the stocking of parts and delay maintenance activities.

It is possible to install a central lubrication system, programmed to meet the needs of the various devices, with feed pipes leading to a central reservoir, or to several reservoirs of different lubricants to be used. The reservoirs can have a low-level warning device and thus save a great deal of time in individual lubrication as well as provide a guarantee that no piece of equipment will be missed. (Also, overlubrication can become as serious a problem as underlubrication.) The expense of central lubrication systems may well be overcome in the years ahead by constantly increasing labor costs.

A system of self-cleaning filtration units for air handling saves the time required for man-handling replacement-type filters, if a practical system is used. It is obvious that this is strictly an economic evaluation, based on the number of such units to be changed, their location, and local wage rates.

Pipelines for drains should be graded with the low point in the direction of flow, otherwise the pipe will be continually dripping or, if provided with a cap, will collect at the cap and spill onto the worker and the floor when it is removed.

Of special importance is the provision of floor drains, with the floors sloped accordingly, to permit degreasing, steam cleaning, pressure washing, and other machine-maintenance activities. Install a curbless floor-type receptor (utility sink) in large mechanical areas; it will have a number of uses.

The use of drip pans, to catch coolants, lubricants, process chemicals, condensation, and other fluids, saves a great deal of time in removing these items from the floor, and also avoids damage to the floor as well as safety hazards from slipperiness. Where a great deal of fluid collection is encountered, a drain can be built into the pan, or the pan can be sloped to a low point which provides quick removal by vacuum. At least, the pan should project far enough from the machine to permit insertion of a vacuum device. (Of course, drip pans should not substitute for an effective maintenance program.)

Shutoff valves should be located to permit convenient maintenance operation and repair of all equipment without the need to trace back to main shutoff valves—which would incapacitate a large amount of machinery or a large area. No valve should be located with the handle pointed downward. Strainers should be located ahead of valves to prevent valve damage through catching foreign material in the seat (these strainers might also avoid major damage to the equipment itself). Double valves should be used on all connections to steam systems that are difficult to shut down for valve maintenance.

Duct systems should be properly supported throughout their length; any sagging will break the joints. Of course, access openings should be provided at all dampers and at important turns to facilitate the cleaning of ducts.

Concerning boilers and boiler rooms:

1. Equip the boiler room with both hot- and cold-water outlets of $1\frac{1}{2}$-in size to be used for washing out boilers. Of course, they will be useful for other activities as well.

2. To prevent the accumulation of dirty water on the floor (as well as to prevent accidents), see that all water glass, water column, water regulators, and blow-down lines feed into a suitable drainage system. Such lines should be open-ended in order to check for leaks.

3. Provide a specific fitting to be installed between the pump and the boiler for injecting boiler-water treatment, assuming that such treatment is desired. Where this is not provided, the passing of the chemical through a pump can do damage to its plastic parts, valves, and the like.

4. Provide the boiler breeching with sufficient cleanout doors of the proper size to make cleaning possible with the least risk of injury and waste of time to personnel. These doors should always be properly sealed with a heat-resistant material such as asbestos.

5. Be sure that the boiler is located so that the tubes may be pulled without either cutting them or removing a wall. The correct condition is most likely to be created where the boiler room is located on a ground, rather than a basement or upper, level.

6. Again, to eliminate the introduction of water within the boiler room, arrange for the venting of boiler safety valves to the outside atmosphere.

7. Where any valve is over 7 ft from the floor, use a sprocket chain on boiler valves (stop valves, feed lines, soot blowers, and so on) to avoid the use of a stepladder.

8. The painting and other maintenance of smokestacks is usually done by specially compensated and equipped workers. Reduce this problem by using stainless steel or brick in the construction of the stack, which eliminates painting and considerably extends the life between replacements. The maintenance of a stack may also be a factor in considering a forced-draft

exhaust, thus permitting a much shorter stack. (Remember the need for monitoring equipment to continuously record air-polluting emissions from stacks, as required by state and federal laws.)

An analysis of the types of periodic and preventive maintenance to be performed on the equipment in question will indicate the need for certain types and sizes of utility outlets. These might include:
1. 120-V electric receptacle
2. 208- or 240-V electric receptacle
3. Compressed air
4. Steam
5. Water
6. Natural gas

HEATING, VENTILATING, AND AIR CONDITIONING

Apart from other considerations, the use of air conditioning (assuming good filtration) considerably decreases the frequency of dusting operations and extends the life of surfaces which can be damaged by dust, such as mineral acoustical tile, fabrics, and other materials. The air motion also reduces staining by cigarette smoke and helps to eliminate objectionable odors.

If the purpose of an air conditioning installation is to provide human comfort, then it is poor economy to skimp on initial cost by providing a borderline or undersized system, or to provide anything other than adequate controls. The result otherwise is continual complaints and service calls on the maintenance department for adjustment and other activities, which never really solve the problem.

A system designed to maintain a slight positive pressure within a building will assure that a minimum of soil-laden air is drawn into the structure.

Be sure not to crowd air conditioning, heating, or ventilating equipment. Ensure accessibility on all sides for repair, replacement of parts, and cleaning, as well as encouraging the all-important inspections. Figure 77 shows one kind of desirable design.

Fig. 77. This long span design provides the structural/electrical platform on the top chord of the truss, and a ceiling/air distribution element on the lower chord. (*H. H. Robertson Company.*)

Controls should be provided—no question about that—but they should not be unnecessarily complex. A standard system, if it can provide sufficient control would be much preferred over one designed specifically for a given installation. Central control panels for smaller systems might be avoided. Valves and controls should be provided by the same manufacturer for best operation.

Stainless steel is the ideal material for chilled-water coils, sump pans, baffles, nozzles, and other items which typically give corrosion problems when exposed to severe conditions. Under normal conditions, copper-tubing coils and galvanized metal are generally considered acceptable.

Too much spinning of the dials of thermostats will damage them and mean extensive adjustments or replacements; they can also be damaged by being struck bodily or with some object.

Thus, thermostats should be located in less-accessible areas (assuming that these areas will provide correct sensitivity) and, when necessary, be further protected with metal guards. Thermostats for temperature control should not be located adjacent to light-dimming devices or other heat-producing devices, as this causes their constant malfunction and readjustment.

Not enough emphasis is placed on the need for isolation valves in piping systems related to heating and air conditioning. Without these strategically located valves, complete draining is required before any part of the system can be properly repaired.

If an air conditioner of the type that exhausts heated air while drawing in fresh-air makeup is used, then the relative location of this air output and input is important. If the fresh-air makeup is drawn immediately from the outside across the coils of a water-heated unit, and the outside temperature has cooled, then this may cause freezing of the coils which can create serious maintenance problems. This can be avoided by locating the fresh-air intake in a more distant place, allowing a "tempering" of the intake air so that it reaches the coils after being mixed with air already in the area, thus considerably reducing the likelihood of coil freezing.

Avoid locating cooling towers where, on windy days, windows or other surfaces might become water spotted. Also avoid placing towers in wells, as air circulation is impeded and tower performance impaired. Finally, do not locate towers where relief or exhaust air continually blows onto them, as this will cause them to dry and result in a shorter life.

The intake duct may be eliminated completely by locating air conditioning equipment on an outside wall; in this case, the louvers themselves become a section of the outside wall. In any case, air-intake or exhaust grills should have bird screens on the outside. Be sure that duct work is equipped with automatic dampering in case of fire.

Be sure time is not wasted in identification by having duct work properly marked. The markings can be made by laminated plastic signs secured by rustproof screws, or, less successfully, with stenciling, located at or near the fan showing the area served by the duct. The identification should indicate revolutions per minute, cubic feet per minute, area served, and so on. Valves should be tagged for identification as well.

A number of organizations find that their greatest heating problems come from the use of a constant-speed exhaust fan in connection with a supply unit that has a system for modulating its fresh-air intake. Thus, at times, the two systems will be working against each other; this should be avoided through the use of a coordinated system.

The pans to catch condensate drip in an air conditioning unit are located under the coils, of course, but they should be of such a design so that they are accessible for cleaning and for the introduction of chemical treatment when necessary. This can be done by having a portion of the pan project from beneath the coils, and a sufficient space between the coil and the pan into which working tools can be inserted. The pans should be constructed of corrosion-resistant materials, or using corrosion-resistant coatings. The drain line from the pan should be of 1-in diameter or larger.

Probably the best type of replacement air conditioning or ventilation filter for most uses would be a permanent metal frame with replaceable disposable pads made of spun glass. Such pads may come in roll form and be cut in size, or precut. The ready-made filters with cardboard binding typically are too expensive, which may result in less frequent change and therefore diminished air conditioning efficiency or the circulation of air containing a good deal of dust. The texture of the filter element—its density and thickness—can be adjusted for the desired conditions in any specific area. Try to standardize on one size. For permanent filtration systems, good results are obtained with the electrostatic or bag type; they do a much better job of dust removal than the replacement filter systems.

Some items on heating:

1. A high-temperature hot-water system for space and process heating can reduce maintenance costs as compared to a steam heating system because of its longer life, the elimination of steam traps, condensate returns, and the minimization of corrosion.

2. A great deal of difficulty has been experienced in some

organizations with hot-water heating coils freezing up on a day when the building was not in use, such as in a public school on a weekend. A hot-water heating system should have proven controls to prevent this occurrence, and should use standard coils so that replacement can be made quickly and easily.

3. Where electric rates are favorable, electric strip heating is preferred over steam fin-tube or wall radiation, in order to reduce maintenance costs, principally in the categories of steam fitting, custodial, sheet metal, and painting. From another viewpoint, hot-water perimeter heating is preferred to electrical and definitely preferred to steam.

CHAPTER NINE

Maintenance Facilities

Location of a maintenance facility is highly important, whether for a separate maintenance structure or for facilities located within a given building. This location has a considerable bearing on the amount of nonproductive time which maintenance workers will experience in getting from the maintenance facility to the place of work.

In campus-type operations, the maintenance building should be rather conveniently located, and is typically adjacent to or part of the power plant, central stores facility, and garage. Yet, it is not infrequently seen that such a facility will be constructed on the periphery of the group of buildings, and sometimes even in a remote location a mile or more distant!

In a given building, including those in a campus-type facility, maintenance facilities should be provided, again as conveniently located as possible. Just as it is economic to strategically locate power, air handling and other equipment in high-rise buildings at varying locations, so it is for maintenance and storage areas. For example, if the power facilities are located every twelve floors to handle the six floors below and above it, it would be logical to consider that same level for maintenance facilities.

Where a single area will be provided in a building of up to about twelve floors high, consideration should be given to providing this on the ground floor level so that it may be adjacent to a loading dock, receiving offices, and the like. If this is not possible, and a basement location is required, try to eliminate, through design, the types of problems typically associated with such areas: ventilation, humidity, and drainage. Where it has been determined to locate maintenance facilities on an upper level, consider the elimination of the basement level of the building altogether.

A separate maintenance structure should be considered where the facility consists of a campus-type layout, such as some types of mental health institutions, a college or university, a research center, or organizations with even more far-flung buildings such as a public school system. Of course, an alternate to a separate maintenance structure would be a portion of a larger building, such as the ground floor of a conveniently located building. A larger facility would contain many of the items on the following list; a smaller one would combine or omit a number of them:

- Maintenance offices
- Maintenance control center
- Locker and rest room facilities
- Garage for vehicles
- Machine shop
- Trade shops or areas, including painting, sheet metal, welding, plumbing, carpentry, electrical, masonry, grounds, custodial, mechanics, air conditioning, refrigeration, electronics, and the like
- In a larger facility, separated specialized facilities, including locksmith, glazier, upholsterer, office machine service, elevator service, storage facilities, and so on

In a relatively smaller organization, maintenance offices might include only two rooms, with trades foremen sharing a couple of desks, while in larger organizations a suite of offices might be required for maintenance management and even for large subordinate departments. Here are some of the requirements for maintenance offices:

1. A primary requirement is a private office for the manager (or subordinate department head where a separate set of offices exist). The office should be sufficiently soundproofed to provide privacy for individual counseling. Normally, a window would look out on the adjacent corridor, but it should be possible to close the window with a shade or blind for further privacy. The office would be equipped with two doors, one to the corridor, and one to the adjacent clerical office. Since the office may be used for conferences, it should contain room not only for the usual furniture, but for several visitor's chairs as well.

2. Adjacent to the manager's office should be an area for clerical and secretarial use. Without this provision, first-line supervisors or even department heads at times become involved in answering telephones, handling inquiries, filling in forms and papers, and otherwise wasting skilled management time. The office would contain furniture for one or more secretaries or combined clerk-secretaries, filing space, several visitor's chairs for salespeople and job applicants, and a wall-mounted magnetic board for organization chart and scheduling purposes. In addition to the door to the corridor and the manager's office, the clerical office might also have doors to an adjacent central storage area and to a supervisor's office. A larger office would be provided for supervisors, preferably with each supervisor having his or her own desk, visitor's chair, and two-drawer filing cabinet; it is not desirable to have supervisors share desks, even if they are working on different shifts. Doors would open to the corridor and the clerical office. For a large facility, a separate lobby and reception area may be indicated.

3. Consideration should be given in the larger maintenance office facilities to providing a combined conference room and lounge area for maintenance supervisory and clerical personnel.

For larger facilities a maintenance training room would be indicated, especially where considerable employee turnover might be anticipated. The basic need and application of such a training area would be for custodial personnel, as they represent by far the largest group of maintenance workers, and also that group whose turnover is normally the most rapid. And although the size of the area would depend on the number of people being trained, it must be at least large enough to both demonstrate and provide practice on various types of maintenance equipment. Thus 500 ft^2 should be considered as a minimum, providing 100 ft^2 each of four types of floors (carpeted, terrazzo, resilient, and concrete) and, in addition, a fifth area of at least 100 ft^2 should be equipped to contain typical furniture, restroom fixtures, and other items to be maintained. And naturally, the area should be equipped with a motion picture screen, folding chairs for training sessions and meetings, controlled lighting, and adequate utilities. To improve the investment in such an area, it could double as a conference or assembly room; in this case, it would probably be desirable to have a number of the fixtures and items of furniture movable so that they may be stored in closets when the room is used for these other purposes.

Where central maintenance offices are located, there should be an adjacent locker room and restroom facilities for maintenance personnel. Consideration should be given to these factors:

Shower facilities

A "break" or lounge area

A first aid or quiet room which could be made a portion of the management offices.

Original equipment of a large maintenance department should include a magnetic staffing or control board. This would be mounted on the wall either in the secretarial or control area of the maintenance offices. Proper symbols can indicate work assignments, personnel absent, the handling of special projects. If a time-clock system is used (and provisions should be made for one, even if it is not initially anticipated), a good location will be the opposite wall in the corridor from the maintenance offices, so that the area is in view from the manager's, clerical, and supervisor's offices. In the same general area a wall-mounted bulletin board should be provided.

A radio-location system for maintenance as well as custodial personnel, such as for group leaders and supervisors, can save a great deal of money in terms of the staff required. Sometimes an adequate system requires consideration of the wiring in the building for it to operate properly, and room for a central station within the maintenance offices. (Such a system is also useful for management and security personnel.)

Making supervisors mobile can improve their effectiveness considerably, and battery-operated carts and scooters are ideal for this purpose. From a design standpoint, this is anticipated through the provision of proper storage space and battery charging facilities, not only in the central storage facility, but in custodial closets and other areas where such scooters might be parked from time to time.

Consider the installation of an alarm system, which could possibly pay for its installation in a single instance, to cover the following:

1. Loss of electric power
2. Air conditioning failure
3. Loss of elevator service
4. Excessive boiler pressure
5. High water or other limit conditions
6. Fire and smoke

Some special considerations for maintenance facilities:

1. Provide a garage for the storage, cleaning, servicing and repair of vehicles of all types, which might include trucks, fork trucks, autos, tractors and scooters. Such an area might also contain a charging station for those items of equipment operated by battery.

2. Building codes vary concerning the amount of flammable material which may be stored inside a building, but they typically limit such quantities to a couple of gallons, which even then must be in proper containers. Where any quantity of flammable materials, such as solvents, fuels, paints, or chemicals which would carry a "red label," are to be kept on hand, a special storage facility should be provided apart from the main building. (See Fig. 78.) This outside structure might consist of a small brick or concrete shed, made inconspicuous with plantings. It might be combined with an enclosure for a large waste compactor or for garbage cans, approached from opposite ends, especially if the common wall is of concrete. Where aesthetics are important, the flammables structure may be located below grade, with proper landscaping to mask the stairway; or better, in a

Fig. 78. At times, building design should contemplate the need for special enclosures for hazardous materials. (*Justrite Company.*)

small hillock to avoid drainage problems and to provide a level entry (similar to a munitions or air-raid bunker).

3. One specific item which should be considered in providing storage space for maintenance equipment is long extension and step ladders. Without a specific location plan in advance, storage becomes a major problem, and ladders are sometimes seen in a corridor or blocking some important utility or service, which of course is a violation of fire and utility codes.

4. It is desirable to have a quantity of each type of surfacing material placed in storage in order to be able to make repairs which will not be noticeable due to change in color, pattern, or texture. This would be particularly true for such items as paint, plastic wall coverings, resilient floor tile, fabrics, and acoustical tile. These materials should be stored in such a way,

where possible, that they will "age" and change coloration under exposure to similar conditions that the original material undergoes. In addition, provide a specific location for spare and replacement parts: piping, conduits, plywood, plastic sheets, window glass, extra flooring, carpet remnants, roofing material, brick, light diffusors, a roll of heavy gauge polyethylene film, etc. Be sure that sufficient storage space is available for such repair items.

5. A special area should be designated for the segregation of items requiring repair or refinishing; this area may actually be a part of the repair area itself.

6. The most common type of maintenance closet is the custodial closet discussed below, but other needs may exist as well, such as for servicing plantings. It is not desirable to use a custodial closet for the storage of electrical and plumbing supplies, and the like; separate closets should be provided, although there may be only one per building.

CUSTODIAL FACILITIES

The central storage area for custodial supplies and equipment would have a loading door opening on to a main corridor or on to a loading dock, plus a personnel door to the maintenance clerical office (Fig. 79). The area might contain these facilities, depending on the size of the operation (a supply room for each building of 50,000 ft^2 or larger would be equipped according to building size):

1. Open storage area for materials normally stored on pallets or skids, such as drums of supplies, paper goods, spare waste receptacles.

2. Steel shelving for the storage of such items as floor machine pads, chemicals in cases, brushes, and mop heads.

3. Racks for the storage of chemicals purchased in drums (although this might be completely eliminated if the determination is made to make purchases in unit containers). Where drums are placed on racks, equipment must be provided for hoisting into position.

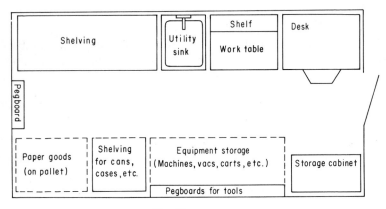

Fig. 79. Typical small central-housekeeping storage area.

4. Locked cabinets for the storage of pilferable items such as aerosol containers, gloves, and the like.

5. A work table for the minor repair and adjustment of equipment and the assembly of newly purchased items.

6. A solution tank for the cleaning of Venetian blinds, egg-crate louvers, and similar devices.

7. A battery-charger setup for charging automatic scrubbing machines, power sweepers, and other battery-operated equipment.

8. A high-pressure washer and/or steam cleaning installation.

9. A floor-type utility receptor, preferably walled in on three sides with ceramic tile.

10. A pegboard or other hanging system for attachments and minor equipment.

11. Racks for the storage of handles and handled items, such as mops and brooms.

12. Closed steel containers, one for soiled and one for clean treated mops and cloths.

13. "Parking" space for custodial carts, floor machines, vacuums, scooters, and ladders.

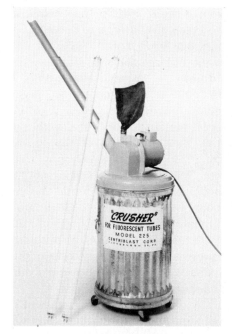

Fig. 80. A fluorescent tube crusher is a useful device for a central maintenance area. (*Centriblast Company.*)

14. A fluorescent tube and bulb crushing machine (Fig. 80).

15. A laundry (combination washer/extractor) for washing and treating dust mops, dust cloths, and other supplies.

16. Lift platforms and collapsible scaffolding, typically used in overhead cleaning and repairs.

17. Initial equipment should always include a drum-head wet vacuum to handle emergency leakages and spillages.

CUSTODIAL CLOSETS

The provision of proper custodial closets is one of the most overlooked items in building design. An insufficient number of them, or their design in an insufficient size, means the wasting of an enormous number of man-hours in going back and forth from a work area to the central supply area. In general, no custodian should have to leave the floor to obtain cleaning supplies, nor walk more than 150 ft to get to the nearest custodial closet on large floors.

Ideally, a custodial closet could be centrally located within the work area assigned to the personnel using it. As a rule of thumb, provide one custodial closet for each 15,000 to 18,000 ft^2 of building floor space, with a minimum of one closet for each floor, regardless of the square footage of that floor. As mentioned in other paragraphs, electrical panels, elevator controls, telephone equipment, roof hatches, and plumbing items should not be located inside the custodial closets. The closets should be located close to an elevator, if possible. The economics of plumbing often indicate the location adjacent to or between restrooms. The location of custodial closets on stair landings represent an impossible working situation; their location under a stairwell invariably provides inadequate storage room and is prohibited by most fire codes.

The closets should be of sufficient size to handle the equipment and supplies of all the personnel who will be using it, as well as containing expendable supplies for a two-week period. These factors assure that custodial personnel will always be within their assigned work area—and therefore much more likely to be performing productive work. In general, a minimum size for custodial closets is 60 ft^2 (although most being constructed these days are less than half this size!); a desirable size would be 70 to 80 ft^2. See Figs. 81 and 82 for designs of a custodial closet of minimum adequacy and an austere closet, adequate for only one worker. Where custodial closets are overdesigned for size, there exists a very real possibility that the area will be appropriated by some other department—such as for an office or for the storage of data processing tapes—and the custodial department will end up with nothing; one way to limit this possibility is to design an area that is more long and narrow rather than square. The custodial closet should contain:

1. A recessed floor-type receptor, without curbs, is desirable. Utility sinks are available both in metal and ceramic construction; in the latter case the curbing should be protected with

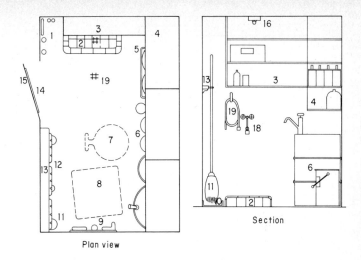

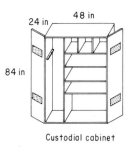

Custodial cabinet

Fig. 81. Custodial closet (minimum). Minimum dimensions: 6 ft wide, 9 ft long, and 8 ft high. Closet to be used only by designated personnel, and to be locked when not in use. If closet is covered, install fan providing 20 air changes per hour. (1) Storage area for hoses, extension wands, pipes, etc.; (2) built-in ceramic tile floor sink, with drain (second choice: wall-mounted utility sink); (3) shelves over utility sink, 9 in deep, 12-in spacing; (4) storage shelving over floor stock, 18 in deep, 12-in spacing; (5) mopping outfit in stored position; (6) floor stock (drums, cans, etc.); (7) floor machine in stored position; (8) vacuum in stored position; (9) accessories, fittings, and tools mounted on pegboard; (10) aluminum or ceramic drip tray; (11) mop in stored position; (12) tri-grip tool holders; (13) 4-in spacer to keep mops away from wall; (14) bulletin board containing instructions, schedules, etc.; (15) 30-in-wide door with louver—location of door interchangeable with accessories pegboard if necessary because of orientation of area; (16) ceiling light providing minimum 40 fc; light should be shielded to prevent damage; (17) floor of ceramic tile or concrete, with floor drain if possible; (18) bibb (threaded) faucet, with brace; (19) length of hose, for washing equipment. A custodial cabinet should be used where there is insufficient space to install a custodial closet.

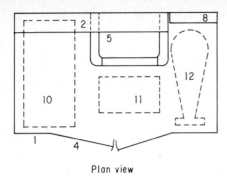

Plan view

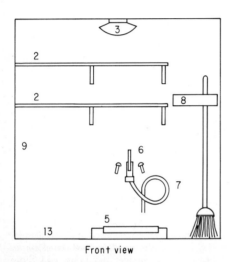

Front view

Fig. 82. Custodial closet (one worker, austere). (1) Dimensions: 8 ft long, 4½ ft deep (36 ft²); (2) shelving 10 in deep, with bracket supports; (3) 100-W lamp, with door-hinge switch; (4) two 30-in doors, pierced for ventilation; (5) utility floor sink, with stainless steel lip cover; note off-center; (6) bibb faucet with support hanger; (7) 4-ft length of hose; (8) tool holder; (9) walls ceramic to 4 ft, painted enamel (including ceiling) above 4 ft; (10) location for custodial cart or waste hamper; (11) location for two-bucket (or three-bucket) mopping outfit; (12) location for floor machine or vacuum; (13) floor—concrete, ceramic, or terrazzo (*not* resilient).

Fig. 83. A number of varieties of floor-mounted receptors are available. (*American Standard Company.*)

a stainless steel lip. The wall mounted utility sinks should be avoided as they are difficult to use, often causing back sprain as workers attempt to lift heavy buckets of water to dump into them; furthermore, they cannot be used to drain solution tanks on wet vacuums or automatic scrubbing machines. The receptor or sink location should be near the door end of the room, preferably (Fig. 83).

2. The faucet over the utility sink should provide both hot and cold water, and of a type that a bucket can be hung from it while being filled. The faucet should be a bibb-type so that a hose can be screwed on it for filling a larger container or fresh water tank in a machine. Naturally, the faucet should be equipped with a siphon breaker so that dirty water will not be drawn into the plumbing system.

3. Shelving for the storage of chemicals and supplies. Wooden shelving is actually more practical than steel shelving in this case, because of the corrosion problem with chemicals,

but stainless steel is ideal, as it avoids painting. (Provide at least 15 ft^2 of shelving, and a 14-in space between shelves 24-in deep.)

4. Hangers for mops should be placed on the wall over the receptor so that wet mops can drain.

5. The custodial closet floor should be preferably of ceramic tile, but other choices are concrete and terrazzo, depending on the economics of construction related to adjacent areas.

6. Custodial closet walls should preferably be ceramic (although they may be plastered and painted with a gloss enamel to about a 5- or 6-ft height) or concrete block if the block is coated with epoxy resin.

7. Ventilation should be adequate, both with an exhaust grille in upper wall or ceiling, and a door that is louvered and/or with an air space at the bottom (provide at least 15 air changes per hour).

8. Illumination should be 50 to 75 fc to provide proper cleaning and care of equipment, and to avoid accidents.

9. An electrical outlet should be provided, as it is often necessary to recharge a battery, and 15 A is required for this; if other activities might occur simultaneously, then the wiring capacity might be 30 A.

10. The doors should be a 36-in lock-type swinging out, with proper identification (most modern departments abhor the word *janitor* and prefer the term *custodial closet* or *service closet* or simply the word *service*). For a custodial closet of shallow depth but considerable width, which might be dictated by space availability, a much better utilization of the space would occur if double doors were provided. But in any case, custodial closet doors should open outward, otherwise a large percentage of the interior space would be lost (in renovation projects, sometimes the reswinging of custodial doors can literally double the amount of available space!).

Custodial cabinets (Fig. 81) are not a proper substitute for custodial closets; however, in a renovation project, they are certainly better than no facilities at all. There are manufacturers of prefabricated cabinets, some even in stainless steel, that merely require hookup to electrical, water, and drain utilities.

The initial outfitting of equipment for custodial closets, such as mopping outfits, floor machines, carts, and vacuums, should be purchased with a white rubber or plastic bumper guard; this prevents the damaging and soiling of walls, doors, and furniture.

CENTRAL VACUUM

The theory and practice of central vacuum systems provide quite a contrast. The theory is that, by having vacuum tubes in the walls with outlets suitable located around the building, one is conveniently able to do vacuuming by simply plugging a hose into one of these outlets, and avoiding numbers of pieces of individual vacuum equipment which must be moved about from place to place. In practice, however, the central vacuum is very rarely a good investment. Most central vacuums that have been installed have since been abandoned, or are simply used periodically for removing the lint from treated dust mops! Most housekeeping managers would much rather have had the annual interest on a rather large investment for the central vacuum to spend on portable vacuum equipment. Here are the drawbacks:

1. The system typically cannot pick up water, and thus portable vacuum units would be needed for this purpose anyhow.
2. Where carpet care is necessary, central vacuums provide only a surface cleaning, and other vacuum units with a pile-lifting feature will be required.
3. Over the years, the pipes become loaded with lint and other soil, reducing their efficiency or clogging them completely. Pipe cleaning is either very difficult or impossible.
4. With "all the eggs in one basket" an inoperative central vacuum leaves no means of providing cleaning.
5. The gaskets in the outlet covers become hardened or out of place, and the resulting air leakage not only is noisy, but also reduces suction.
6. The long lengths of hose necessary to provide vacuuming service in all areas causes considerable damage to furniture, door frames, and other surfaces by being dragged about; workers

often refuse to work with these hoses because of the difficulty of handling them.

One good application of a central vacuum is an industrial "superclean room" where a requirement is that the exhaust be exterior to the room; other possibilities include a hospital operating room and a data processing area. On the other hand, in all of these cases, special portable wet-vacuums are available, with a double filtration system, which exhausts particles only below $\frac{1}{2}$ μm in size; this is a very small particle, considering that the typical bacterium dimension is 1 μm.

Suggested Reference Literature

J. H. Callender (ed.), *Time-Saver Standards for Architectural Design Data,* 5th ed., McGraw-Hill Book Company, New York, 1974.

H. S. Conover, *Grounds Maintenance Handbook,* 2d ed., McGraw-Hill Book Company, New York, 1958.

Robert H. Emerick, *Heating Handbook,* McGraw-Hill Book Company, New York, 1964.

Edwin B. Feldman, *Housekeeping Handbook for Institutions, Business and Industry,* Frederick Fell, Inc., New York, 1969.

B. T. Lewis and J. P. Marron, *Facilities and Plant Engineering Handbook,* McGraw-Hill Book Company, New York, 1973.

Frederick S. Merritt (ed.), *Building Construction Handbook,* 3d ed., McGraw-Hill Book Company, New York, 1975.

L. C. Morrow, *Maintenance Engineering Handbook,* 2d ed., McGraw-Hill Book Company, New York, 1966.

John H. Watt and Wilford Summers (eds.), *NFPA Handbook of the National Electrical Code,* 4th ed., McGraw-Hill Book Company, New York, 1975.

George O. Weber (ed.), *A Basic Manual for Physical Plant Administration,* Association of Physical Plant Administrators of Universities and Colleges, Washington, D.C., 1974.

Checklist

ACTIVITIES DURING CONSTRUCTION

See that the future maintenance manager is on hand during construction.

Consider a salaried inspector to supplement architectural inspections.

Prepare for sophisticated maintenance if sophisticated systems are used.

Take advantage of standardization, but review the standards regularly.

The Construction Contractor

Check the construction contract for methods specified to limit soil creation.

Obtain a careful preoccupancy cleaning.

If temporary heating is required, see that it does not produce soot.

See to the adequate use of building paper.

Trades Work

Use machine troweling for concrete floors, including subfloors.

Check the details concerning application of resilient flooring; they are important.

Use a vacuum attachment device for terrazzo and concrete grinding.

Use disposable air conditioning filters during construction.

Mount restroom fixtures at the proper height.

Be certain that floor drains are indeed at the lowest point of the floor.

Avoid painting wood sash and screens closed.

Before applying coverings, remove any pencil marks or other foreign material on walls.

Replace marked or soiled ceiling tile.

Exterior Considerations

Give adequate attention to the details of backfill material.

Be sure that soil is compacted properly.

Pay attention to grounds care before and during construction.

Protect plantings that are to be saved.

Protect walkways or paving from damage.

Remove soil from streets or walks to prevent its being tracked into buildings.

Protect catch basins and storm drains from water and concrete spoil.

Protect any surface which is susceptible to damage during exterior cleaning.

As with floor drains, be sure roof drains are at low points.

Custodial Maintenance

Provide adequate staff for custodial maintenance during construction.

Follow the principle of isolation for construction soils:
- a. Provide airlocks in hallways
- b. Cover some items to protect them from dust accumulation
- c. Use mats and runners to trap soils
- d. Check air conditioning filters regularly
- e. Use "off limits" signs
- f. Control sight-seeing
- g. Protect equipment or machinery near the construction area

Control the soil load in occupied areas.
- a. Wet clean floors regularly
- b. Vacuum walls, trim, and furniture
- c. Thoroughly wash walls and trim after construction
- d. Use a protective coating of floor wax
- e. Give special attention to connecting areas
- f. Plan for more frequent cleaning of stairs and elevators

Final Activities

See to the accuracy of as-built drawings and their protection.

Also see to the provision and protection of operating and maintenance instructions.

For optimum operations, prepare the maintenance manual before occupancy.

Keep the premises free from accumulations of waste and tramp materials at all times.

Properly balance hydraulic and ventilation systems.

Test and demonstrate all systems before accepting them.

Try to avoid the pitfalls of "beneficial occupancy."

Consider specialists for planning and actually making the move into the building.

Use a promotional system for obtaining cooperation from building users.

FROM THE GROUND UP

Below Ground

Consider soil sterilization where termite infestation can become a serious problem.

Landscaping

Provide soil suitable for lawn growth.

In difficult situations, use low-maintenance ground cover.

Avoid steep slopes for grassed areas; 2 percent is ideal.

Consider a built-in lawn sprinkler system.

Avoid small or irregularly shaped grass plots.

Provide concrete mowing strips to eliminate hand trimming.

Rely on proper walkway design to avoid lawn damage.

Consider man-made turf substitutes in problem areas.

Use plantings which are indigenous to the area.

Eliminate flower gardens if possible.

Avoid overplanting trees and shrubs.

Provide a 12-in unplanted space around trees to avoid mowing damage.

Set trees and shrubs back at least 6 ft from curbs.

Where the building design includes a courtyard, provide adequate entry for all maintenance equipment to be used in it.

Paving

Apply concrete or asphalt paving over a well-constructed base, coated with sealers.

Do not use asphalt on a grade steeper than 10 percent.

Consider the size of grounds-care equipment in the design of walkways, roads, and so on.

Provide diagonal as well as rectangular walkways between buildings.

Install a 5-ft minimum radius at walkway intersections.

Give walkways a 2 percent transverse slope, or a crown for drainage.

Avoid feature strips or various materials in concrete walks.

Avoid built-ins in walks and drives; they are difficult to sweep or snowplow.

Barricade any paved areas that cannot support vehicles.

Set columns or other construction near driveways well back from the curb.

Use guards to protect the corners of walls adjacent to driveways.

In asphalt truck-parking areas, provide concrete strips to support steel wheels.

Install a storm drain in loading and parking areas.

Grade loading and parking areas away from the building in the event the drain becomes clogged.

Consider built-in snow melting equipment, especially for walkways.

Use granite curbing where snow plows are used. It is most successful.

Provide ramps at curbs for wheelchairs and other such equipment.

External Features

Provide a groundskeeping closet on the exterior.

Use a pea-gravel splash area to prevent mud splatter on lower outside walls.

Carefully consider the many factors involved in bulk waste containers.

Light parking areas to reduce vandalism and damage.

Where two or three steps are planned, replace them with ramps.

Provide a well-designed loading dock.

Install a concrete strip under fencing to avoid mowing and litter problems.

Eliminate exterior drinking fountains where possible.

Soil Entrapment

Provide a combined system of soil entrapment to eliminate the trackage of materials into a building, considering the following possibilities:

- A covered walkway
- A roughened walking surface
- A grating with catch pan (of primary importance)
- Areas with special carpet for this purpose
- Matting
- Runners

Exterior Building Surfaces

Concentrate on using maintainable surfaces such as polished stone, stainless steel, and glass for exterior surfaces.

Also concentrate on exterior walls of brick, cast stone, natural stone, and other permanent materials. Avoid wood.

Avoid using porous stone, or at least use rustproof fasteners.

Avoid horizontal surfaces on precast concrete, plastic, or metal panels.

Install compression layers on masonry walls to avoid damage caused by shrinkage.

Avoid designs that create natural ladders.

Avoid steel external stairways.

Do not use signs made of individual letters affixed directly to a wall.

Try to avoid irregular building configurations which both provide concealment for vandals and increase the surface to be maintained.

Roof

Do not use a flat roof.

Leave built-up roofs flat; avoid aggregates where possible.

Consider plastic and aluminum for industrial roofs.

Provide adequate roof drains.

Be sure guttering is of ample size.

Where corrosive vapors are present, use gutters and downspouts of resistant material.

Install permanent ladders for roof access if elevators are not available.

Use roof walkways or tread boards for inspection and maintenance purposes.

Consider the values of a roof overhang.

See that canopies have drains.

Use a steel sleeve or boot to protect downspouts where they may become damaged.

Consider interior downspouts in cold weather climates.

Provide leaf cables for exterior downspouts in cold weather areas.

Avoid ferrous materials for flashing and related items.

Provide a pitch pocket where wires or bars penetrate the roof.

Where columns and other structural members penetrate the roof, use a pedestal flashed with a cap.

Avoid close spacings on items penetrating the roof.

Use a deflector hood or skirt for ventilator pipes or similar items piercing the roof, especially if the roof is not well sloped.

Provide lead caps on horizontal top joints of stone copings.

Avoid flat surfaces on tops of walls and parapets.

Try to eliminate parapets and other decorative roof features (except for fire control).

Protect roof-mounted mechanical equipment from the elements.

Consider relocating mechanical equipment that might be mounted on the roof.

Consider the advantages of high-performance roof insulation.

Bird Control

Try to eliminate bird problems by avoiding surfaces on which they can alight or nest.

Consider the installation of various devices that prevent birds from alighting.

FLOORS, ELEVATORS, AND STAIRS

When replacing worn or damaged floors, use a contrasting rectangular color.

For conductive floors, use terrazzo. It is the best investment.

Install an automatic resistance-testing system for conductive floors.

Where pedestal floors are used, specify plastic only.

Where floors are continually wet, provide trenches or drains.

On wet floors provide integrally poured concrete curbs and sills.

A cove from floor to wall is practical, but should be no higher than 6 in.

Where aisle markers are needed, place them permanently.

Avoid open-slot expansion joints.

Do not pour concrete slabs in direct contact with masonry walls.

Carpeting

Remember that carpet has many advantages for specific situations, but is not universally desirable.

Avoid solid colors.

Avoid cut or deep pile.

For carpets, consider:
- Dense, low pile for wheelchair or cart traffic
- Three-color shag for light-use areas

- Large ball-type casters for office areas
- Continuous surfaces; do not mix carpet and other flooring
- Flame characteristics of various types of carpeting

Remember that ideally carpet should have these characteristics:
- Four-color tweed pile
- Tight-loop pile
- Continuous synthetic filaments
- Pile height of approximately ¼ in
- Impermeable backing membrane

Flooring Types

Avoid asphalt tile, which is a poor investment although initial cost is low.

Consider ceramic tile, which is highly durable and has a generally trouble-free surface. (Use dark grout.)

Try concrete for industrial and other uses where appearance is not a prime factor, but remember that it should be densified or sealed.

Remember that cork's acoustical qualities are better provided by carpet.

Be cautious with homogeneous vinyl, which is both expensive and tends to show subfloor irregularities.

Be cautious with impregnated wood, which has some of the same drawbacks as natural wood although it is more durable.

Avoid linoleum, as after a time it develops a poor, irregular appearance.

Be cautious with marble, as its porous structure is easily stained, and it is easily damaged by harsh chemicals.

Consider plastic laminates, which are good for some problem situations, but can give irregular performance based on installation.

Consider plastic tile, which is expensive but is generally maintenance-free.

Also consider rubber tile, which is also expensive. (Vinyl asbestos is usually preferred.)

Remember that stone is durable if well set; grout must be sealed; but appearance may deteriorate.

Try terrazzo, one of the most durable and attractive of all surfaces, but avoid white terrazzo; terrazzo must be sealed.

Avoid travertine except where crevices are filled with epoxy resin.

Use vinyl asbestos, generally the best investment for most floors, with competitors being concrete, carpet, and terrazzo.

Remember that wood is susceptible to damage by water and mechanical scratching, and should be sealed.

Elevators

Place waste receptacles and cigarette urns between every two elevators.

See that at least one elevator serves all floors, including basement and roof.

Be sure of elevator size and capacity for moving equipment.

Remember these best elevator surfaces:
- Doors: plastic-laminate
- Floor: carpet (as a soil trap)
- Walls: plastic-laminate
- Ceiling: plastic panel

Make sure that every elevator contains an electrical outlet.

Make sure that every elevator contains a wall-mounted cigarette urn/waste-receptacle unit.

Provide a large freight elevator, preferably one that goes to the roof.

Locate the freight elevator for easy access from the exterior of the building.

Be sure the elevator shaft contains proper lighting for maintenance.

Size and equip the elevator control room for maintenance.

Stairs

Avoid resilient tile, bluestone, slate or marble for stair treads and landings.

Use good surfaces—precast terrazzo, stone, and formed synthetic materials.

Use carpeted landings, which are beneficial as a soil trap.

Remember that stairwell walls should be durable and not easily marked, such as natural brick.

Be sure that handrails are simply designed, continuous, and well attached.

Do not locate lights over stairs; also, they should be replaceable from the landing.

See that landings contain a wall-mounted cigarette urn/waste-receptacle unit.

Provide for an electric receptacle on each stair landing.

Use a ramp rather than a stairway of three or less risers.

WALLS AND CEILINGS

Wall Coverings

For vinyl coverings, use heavy material that is not sharply embossed.

Avoid burlap, natural grass paper, and felt.

Use epoxy-resin coatings for a smooth, durable surface.

Avoid flat paints, which mark easily and are difficult if not impossible to wash.

Avoid specially mixed paint colors.

See that art reproductions are fully encased in plastic.

Where many wall hangings are to be used, provide a hanging strip.

Install a high electric outlet for lighted paintings.

Wall Materials

Use glazed tile for problem areas such as those for food processing, health care, and so on.

If wood is to be used, specify plastic-protected varieties.

For plastered walls, locate casing beads where cracking would otherwise occur.

Finish plastered corners to a quarter-round metal.

For gypsum wallboard, use two ⅜-in laminated panels for best results.

Concerning concrete block walls, remember:
- To avoid raw concrete; painted is better, but epoxy-coated concrete block is most desirable.
- That glazed block is helpful, but the grout is still unprotected.
- To use rounded blocks on outside corners.
- To use a molded cove base.

For the most successful elevator wall, use plastic laminate.

Avoid wood bases for walls; use vinyl or rubber cove bases.

Remember that carpet baseboards are practical for carpeted floors.

Give special attention to lobby walls.

Design For Walls

Limit graffiti by using both materials that are difficult to mark and good illumination.

Avoid wall-design features that break up the smooth, uninterrupted surface.

Use a minimum width of 8 ft for corridors in order to avoid wall damage.

Use rounded corners in heavily trafficked intersections.

Use movable panels or design according to the "office-landscaping" concept where three or more moves are to be expected in the life of the building.

Consider stainless steel corner guards for walls and columns where equipment or furniture will be moved regularly.

Enclose fire extinguishers in recessed cabinets.

Remember that natural surfaces, such as stone or brick, may eliminate much maintenance.

Acoustical Ceilings

Do not use soft, blown-on mineral materials; they simply cannot be cleaned.

Do not glue or nail acoustic tile to ceiling surfaces; use lift-out tiles.

If mineral acoustic tile is to be used, remember that the plastic-faced variety is washable.

Provide air relief for lay-in tiles in foyers and entranceways.

Remember that aluminum metal-pan ceilings are the most durable acoustic materials.

Use a modular, integrated ceiling to avoid numerous problems, such as with rings around diffusors.

Where diffusors are used, be sure the adjacent surface is smooth tile or metal to provide for easier cleaning.

Ceiling Design

Be sure that ceiling heights are adequate for even the lowest ceiling, such as often is found in corridors.

Where plastic ceilings are used, be sure they are removable.

To make plastered ceilings practical, provide acoustical treatment with carpets, drapes, plantings.

Avoid any textured finish on dry wall or other wall surfaces.

Consider eliminating some ceilings altogether by "painting out" areas above the illuminated level.

Try to eliminate the installation of any equipment in overhead areas.

Avoid the problems associated with skylights by using artificial lighting.

FURNITURE, FIXTURES, AND FENESTRATION

Keep furniture styles as simple and uncluttered as possible.

Avoid floor surface moldings to cover electric wires.

Arrange for straight, uncluttered aisles.

Avoid overcongested furniture areas.

Inside a building, use synthetic plantings rather than natural flowers.

Where buildings are only partly used at certain times of day, provide means of locking off unused parts.

Note that the owner's approach to smoking regulations will determine a number of features.

Where rooms and space are rented, install a room-notification system.

Provide a cloak closet—next to the elevators—for visitors.

Waste Collection and Disposal

Consider the need for space and electrical requirements if bottle- or can-crushing equipment is provided.

Garbage-storage areas should be equipped with water outlet and floor drain.

Locate trash and garbage collection and storage areas in accessible spots.

Where the owner is required to identify and store types of waste, provide two rooms for this purpose.

Locate and size waste receptacles according to waste generation.

Vending Machines

In campus-type areas, consider a central vending structure.

Avoid scattering vending areas in any building; centralize this equipment.

Furniture

Wall-mount as much furniture as possible.

Require that furniture surfaces be cleanable and that they resist marking and scratching.

Use plastic laminates for desk and table surfaces.

Require a toe space under counters, displays, cabinets, and other fixtures.

Remember these items for furniture maintainability:
- Minimize the number of legs.
- Consider the use of collapsible tables.
- Use stainless steel glides with a diameter of at least $1\frac{1}{4}$ in.
- Avoid the use of grilles and screens.
- Avoid unprotected wood.
- Provide deep undercuts for finger pulls in cabinets and drawers.
- Use concealed rather than exposed hinges.
- Avoid upholstered or fabric-covered headboards.
- Use anodized aluminum or wooden bases for lamps that are susceptible to rusting.
- Use recessed ceiling tracks for cubicle curtains.
- For files, use recessed cabinets or floor-to-ceiling cabinets.

Fixtures

Protect fixtures by locating them in recessed areas or by using curbs and rails.

Be aware of the several design features that affect mobile carts.

Use only large, heavy-duty casters on mobile equipment.

Wall-mount drinking fountains; recess them as far as possible.

Avoid painted metal fountains; use ceramic or fiber glass.

In wet areas, place fixtures on elevated bases to avoid rust.

Where fixtures may flood or discharge water, avoid locating them overhead.

Provide adequate ventilation for refrigeration equipment.

In designing food-handling or food-processing equipment, consult the National Sanitation Foundation.

Provide slanting rather than horizontal tops on such items as sills, lockers, radiator covers, baseboards.

Windows

Attempt to eliminate windows altogether.

If windows must be used, consider fixed rather than operating ones.

Remember that pivoting windows can be washed from inside a room.

Avoid any internal treatment that will make pivoting windows inoperable.

Avoid the overuse of glass.

Flat glass is preferable to corrugated, embossed, or ribbed glass.

Prevent damage to large panes of vertical glass by using adequate bars or marking.

Avoid glass block.

Do not design window ledges that are near floor level in such a way that they can be sat or stood on.

Consider glazing with plastic panes rather than glass where breakage is a problem.

In industrial-type situations, use corrugated fiber glass for roof lighting.

Rabbet design should provide handling and setting of glass without damage to the rabbet or the glass.

Use mullions of stainless steel for long life and little maintenance.

Consider using the mullions as guide tracks for an automatic window washing system.

Design a fixed track around the perimeter of the roof for supporting window washing and other maintenance staging.

If vertical blinds are used, specify heavy-duty hardware.

Consider using double-pane windows with a Venetian blind between. They provide good service if you choose the design with care.

Use fiber glass draperies or curtains.

Remember these window design features:
- Choose aluminum windows over wood or steel.
- Avoid setting exterior louvers or decorative masonry that will interfere with window washing.
- If windows are to be washed manually, provide suitable hooks.
- If windows are flush with exterior walls, provide a means of conducting water around the window.
- Consider tinted, heat-reducing, glare-reducing glass.

Doors

Where possible, use an arch or opening rather than a door.

Be sure doors are large enough to permit movement of equipment and supplies.

In door selection, consider the possibility of change of floor surface height.

Concerning exterior doors, remember:
- In severe-weather areas, protect doors with vestibules, canopies, and so on.
- Construct outside doors of aluminum or stainless steel.
- Design door pulls to avoid a lever action.
- Avoid center posts, or make them removable.
- Remember the use of floor grating in door design.
- Frame glass doors in aluminum or stainless steel.
- Provide a slope away from the door to prevent water backup and icy spots.

Door Design

For best economy, use heavy-duty hardware throughout.

Where through bolts are used, provide spacer sleeves to prevent door collapsing.

Use door closers on the hinged side, thus avoiding a bracket.

Avoid floor-type closers.

Standardize hardware types throughout.

Do not use floor-mounted door stops.

For push plates and kick plates, use plastic laminates.

Use a hardware system that resists tampering and loosening through shock or vibration.

Remember these details on door design:
- See that the door is flush or uses the simplest form of trim.
- Avoid the narrow-line aluminum doors.
- Use solid-core doors where damage and sound control is a factor.
- See that all door frames are of metal.
- Provide protection or warning for full-length glass doors.
- Avoid the use of louvered doors for decorative purposes.
- Use rubber swinging doors in industrial warehouse areas.
- Equip doors to mechanical and other such rooms with self-locking locks and free knobs on the inside.

Establish a good key-control system and install a key cabinet.

Remember that complete building design requires a practical door numbering system.

RESTROOMS, PLUMBING, AND PIPING

The restroom layout should minimize walking and the dripping of water.

Consider gang facilities for some installations.

In schools, do not separate faculty from student facilities.

Remember that sometimes rapid conversion from men's to women's restrooms is required.

Use doorless entrances where possible.

Design restrooms to permit spray cleaning.

Control odors by surface design and ventilation.

Limit graffiti by using surfaces difficult to mark, bright illumination, and any other reasonable means.

Restroom Surfaces

Use unglazed ceramic tile for the best restroom floor.

Use tile grout of a dark color.

See that a floor drain is installed.

Use glazed ceramic tile for the proper wall surface.

As with floors, use a dark grout for wall tile.

Hang stall partitions from walls or ceilings.

Equip at least one stall for wheelchair use.

Use double tissue dispensers and other restroom accessories.

See that partitions are hung well.

Washroom Fixtures

Consider all the factors involved in a lavatory selection.

For urinals, use the wall-mounted, flooded open-throat type.

Consider a timed flushing device for urinals.

For urinals and toilets, use flush valves, mounted 36 to 42 in above the floor.

In toilet-seat selection, try to avoid a seat cover; use an open-front elongated seat; provide a self-raising seat; use solid plastic; avoid black bumpers.

For circular wash fountains, make fiber glass your first choice, stainless steel your second. (Avoid terrazzo.)

If a custodial sink is not provided, install a wall-mounted bibb-type faucet.

Specify built-in antislip features for tubs and showers.

Restroom Equipment

Use recessed, multipurpose units of large capacity.

In women's restrooms, use a wall-mounted recessed napkin dispenser; use wall-mounted receptors in each stall.

Do not put mirrors over lavatories.

Avoid glass soap containers.

Install a soap dispenser over each lavatory.

Consider a central distribution system for liquid soap.

Specify paper towels dispensed from rolls; use electric hand dryers in public locations.

Mount lockers on a solid base with a ceramic cove.

Slant the tops of lockers to avoid soil collection.

Plumbing

Permanent drawings should show location of all plumbing systems, including underground.

Use an identification system for plumbing and piping systems.

Use a color-coding system for piping.

Standardize fixtures as much as possible.

Provide adequate space for working on plumbing and piping.

Locate piping so it does not obstruct openings.

Protect maintenance workers from hot piping or dripping liquids.

If you use floor drains, be sure they are adequately sized.

Use red brass for water piping.

Use plugged tees rather than elbows to provide cleanout points.

Be sure to provide enough valving so that systems can be closed down in sections.

Always provide siphon breakers for bibb-type faucets.

Use pipe unions at each valve for ready removal.

Standardize valves, and keep spare parts on hand.

Use chains or extension handles to control overhead valves from floor level.

In lavatories, valve each hot and cold riser.

In cold-weather areas, protect outside faucets and bibbs.

Be sure that hydrants are cleanable without disturbing the barrel or casing valve when shut.

Consider the use of sprinkler stoppers to control water flow where fire protection is not required.

Specify corrosion-resistant drip pans where leakage is a problem.

Equip laundry chutes and waste chutes with floor drain and cleaning spray head.

Consider the installation of water softening equipment.

Install piping to permit steam condensate to drain by gravity.

Provide a metal sleeve for pipe that passes through walls.

Use a metal saddle to protect pipe supported by rollers.

Avoid damage to pipes by using antifriction piping supports on long runs.

For future use, provide plug stub-tees.

Where pipes or valves are subjected to temperatures over 190°, use a heat-resisting paint.

TRADES MAINTENANCE

Structural

Provide ready access for piping, wiring, and conduit.

Consider specific utility spaces, either above, below, or vertically through the space served.

Provide utility tunnels for clusters of buildings.

In industrial facilities, use service galleries.

Install permanent ladders for frequent service conditions.

Provide a lifting device for removal and repair of equipment.

Use adequate expansion joints.

Avoid trussed structural members.

Use a beveled filling between flanges of vertical columns to avoid collection of soils.

Use hinged or removable access panels for conveyor systems or similar equipment.

Electrical

Most important: provide adequate capacity.

Install empty conduit in some situations to handle future requirements.

Consider the likelihood that building areas will be upgraded and provide sufficient electrical requirements initially.

In changing situations, use rack, tray, or channel construction for wire supports.

Again, in changing situations, consider bus bars.

Do not space electrical outlets at excessive intervals.

In restrooms, provide outlets for everything from electric shaving to powered cleaning equipment.

Avoid canopy switches, pull-chain operation, and light-fixture outlets.

Do not overlook facilities for battery charging for various types of equipment.

Avoid many problems by eliminating floor-mounted receptacles.

Do not permit running water and drain lines to pass through electrical equipment areas.

Provide work room around electrical equipment for easy, safe maintenance.

Protect electric gear from physical damage.

Provide specific enclosure for high-voltage equipment.

See that electric gear areas are well ventilated.

Be sure that adequate insulation is provided to protect maintenance workers.

Be sure that electric vaults are accessible from building exteriors.

See that indoor electric vaults contain a floor drain.

Use identification systems for electric wiring and equipment.

Lighting

Be sure that lighting fixtures are accessible.

Never place lighting fixtures over stairs. Put them over the landings.

Use rough-service lamps in problem areas.

Provide central lighting controls for each floor.

Allow for the possibility of alternate row lighting.

Where possible, use standard voltage for lighting.

In areas normally dimly lit, provide alternate lighting for maintenance.

Do not locate accent lights in hard-to-reach areas.

Use automatic flag-pole lighting to avoid repeated flag-raising ceremonies where these are expensive.

Consider exterior floodlighting to minimize vandalism.

Where a number of long fluorescent tubes are used, provide a tube crusher.

Select fixture designs that do not collect dirt.

Use low-tempeature lamps where conditions warrant.

Try to avoid suspended fixtures.

Use high-impact plastic rather than glass globes.

Where dust is a problem, consider ventilated fixtures.

Avoid egg-crate diffusors; use solid lens instead.

Specify an interior fixture surface that is easy to clean.

Mechanical

Consider the use of scale models to eliminate problems.

Be sure that drawings show mechanical equipment in open, or extended, positions to assure clarity and accessibility.

Use built-in hoisting equipment for heavy machine maintenance.

Consider the installation of an overhead crane.

Provide separate space secured from the occupants of a building for major mechanical and electrical equipment.

Provide sufficient illumination for machine maintenance.

See that the correct information is available initially for future maintenance.

Identify equipment with laminated plastic signs.

Consider a central information-gathering panel board.

Minimize the need for special tools and devices.

Consider a central lubrication system.

Use self-cleaning filtration units for air-handling systems where possible.

Grade pipelines for drains with the low point in the direction of flow.

Use floor drains in areas for mechanical maintenance where spillages may be expected.

Provide drip pans to catch coolants, lubricants, process chemicals, and other fluids.

Locate shutoff valves for convenient maintenance.

Be sure that duct systems are properly supported throughout their length.

Remember the following boiler maintenance items:
- Provide $1\frac{1}{2}$-in hot- and cold-water outlets in the boiler room.
- Feed all blow-down and other fluid lines to a suitable drainage system.

- Provide a specific fitting for the introduction of boiler water treatment.
- Use sufficient cleanout doors for the boiler breeching.
- Locate the boiler so the tubes may be pulled easily.
- Vent safety valves to the outside.
- For valves that can not be reached, use a sprocket chain.
- Use smoke stacks that do not require painting.

Heating, Ventilating, and Air Conditioning

Remember the general value of air conditioning in reducing airborne soils.

Avoid borderline or undersized systems.

Remember that the system should maintain a slight positive pressure within the building.

Provide adequate accessibility on all sides of the units.

Do not make air conditioning controls unnecessarily complex; use standard systems if possible.

Remember that stainless steel is the ideal material for water coils, pans, and so forth.

Locate thermostats in less accessible areas; protect them with guards where necessary.

Use isolation valves so that the entire system need not be shut down for maintenance.

Locate fresh-air intake so as to provide a tempering of intake air.

Locate cooling towers so that windows and other surfaces are not water spotted.

Plan the location of the intake duct carefully.

Provide permanent identification signs for equipment and ducts.

Use a coordinated system of exhaust fan and fresh-air intake.

Catch pans should be accessible for cleaning and emptying.

Catch pans should be constructed of corrosion-resistant material.

Select the filter medium and system carefully.

Consider the advantages of a high-temperature hot-water system, rather than steam.

Provide controls to eliminate freezing of the hot water heating system.

Consider electric strip heating where electric rates are favorable.

MAINTENANCE FACILITIES

Provide a centralized location for maintenance facilities to limit travel time.

In a given building, remember that maintenance facilities are best located on a ground floor rather than on a basement or roof level.

In campus-type facilities, consider a separate maintenance structure containing these features:

- Maintenance offices
- Maintenance control center
- Locker and restroom facilities
- Garage for vehicles
- Machine shop
- Specialized trade and shop facilities

In maintenance offices, include a private office for the manager as well as offices for the clerical and secretarial help.

For larger facilities, include a maintenance training room.

Locate locker and restroom facilities next to maintenance offices.

Consider a staffing or control board, which can be very useful.

Provide for a time-clock system even if it is not initially anticipated.

Consider the possible use of a radio-location system on maintenance design.

Provide for battery charging of mobile equipment.

An alarm system, reporting on numerous features, can even save the entire facility.

Consider these special features for maintenance facilities:
- For flammables in other than small quantities, provide a
- Provide a garage for vehicle maintenance.
special storage-space.
- Provide for ladder storage.
- Set aside storage space for replacement quantities of building-surface materials.
- Provide a special space for segregation of items requiring repair or refinishing.

Custodial Facilities

Provide for central storage of custodial supplies and equipment (seventeen items are listed in the text).

Custodial Closet

Provide a custodial closet on each floor.

Remember that a custodian should not have to walk more than 150 ft to reach a custodial closet.

Typically, provide one custodial closet for each 15,000 to 18,000 ft^2 of floor space.

Centrally locate custodial closets (usually next to restrooms).

Do not locate closets on stair landings.

Closets should be 70 to 80 ft^2; a minimum is 60 ft^2.

Remember that a number of factors are involved in custodial closet design (the text lists ten).

Emphasize that custodial cabinets are not a substitute for closets; they are a last resort.

Specify that custodial equipment must have white rubber bumper guards.

Central Vacuum

Except for a few very specialized situations, central vacuum systems are not usually a good investment.

Index

AABC (Associated Air Balance Council), 31
Access panels, 159, 161, 172
   illustrated, 160
Acoustic ceiling, 2, 22, 95–101, 174
Acoustic door, illustrated, 5
Acrylic/wood floor, 70
Air conditioning, 22, 100, 119, 126, 127, 165, 174–177
Air conditioning controls, 174
Air-conditioning-equipment rooms, 155
Air curtain, illustrated, 48
Air diffusor, 100, 101
Air lock, 27, 129
Air pollution, 174
Airports (see Transportation terminals)
Aisle markers, 64
Aisles (see Corridors)
Alarms, 162, 171, 183
   fire, 31, 157
Aluminum, 49, 56, 57, 59, 82, 117, 130, 137, 147, 169
Aluminum-pan ceiling, 99
American National Standards Institute, 152
Anemostats, 100, 101

Angles, 11
Appliances, electric, 161
Architects, 4, 5, 13, 18, 29
Arson, 160
Art reproductions, 87
Artificial turf, 36
As-built drawings, 29
Asphalt paving, 38
Asphalt tile, 67, 80
Associated Air Balance Council (AABC), 31
Atmosphere, polluted, damage caused by, 15
Automatic scrubbers, 28, 130, 164, 186, 191
Automatic window cleaner, 124
   illustrated, 125

Backfill, 23
Bacteria control, 47, 72, 152, 194
Baked enamel, illustrated, 90
Balancing, system, 31
Baseboard, 93, 119
Basements, 180
Bathmat, 146
Bathtub (see Tubs)
Battery charging, 164, 183, 186, 192

Beds, 116, 128
Beneficial occupancy of incompleted buildings, 31
Bibb-type faucet, 146, 155, 191
  illustrated, 189, 190
Bird control, 60, 176
Blown mineral ceiling, 96
Boiler rooms, 156, 173
Boiler-water treatment, 173
Bolts, 142
Bottle crusher, 109
Bradley basin, 136, 145
  illustrated, 146
Brass, 82, 131
  red, for water piping, 155
Brick, 74, 82, 89, 173
Bronze elevator doors, 78
Building code, 138, 183
Building manager, 13, 26
Building paper, 20
Built-in furniture, 112
Built-up roof, 54
Bulletin boards, 87, 182
  illustrated, 189
Bumpers, 154, 164, 193
Bus ducts, 162

Cabinets, storage, 186
Cafeterias, furniture for, 115
Campus:
  central vending facilities on, 111
  maintenance building on, 179, 180
Can crusher, 109
Canopy, 58, 128, 130
Carpet (see Floors, carpet)
Carpet squares, 7, 79
Carts, 94, 117, 122
  illustrated, 96
Casters, 118
Catch basin, 24
Catwalks, 160
Ceiling design, 101–104, 161, 165
Ceilings:
  acoustic, 2, 22, 95–101, 174
  aluminum-pan, 99
  blown mineral, 96
  false, 27
  integrated, 100
  suspended, 98

Central lubrication, 171
Central soap system, 149
Central stores, 179, 183, 188
Central vacuum, 193–194
Ceramic tile, 9, 25, 67, 78, 89, 91, 93, 136, 138, 140, 192
  illustrated, 88
Chases:
  plumbing, 153
  utility, 160
Chemical cleaning, 25
Chemical plant, 118, 164–166
Chemical storage, 191
Chemical treatment (air conditioning), 177
Chilled-water system, 118
Chutes, laundry, 157
Cigarette urns, 78, 79, 83, 107, 143
Circuit breakers, 167
Circular washbasin, 136, 145
  illustrated, 146
Cleaning, 161, 164, 165, 167, 178
  contract, 26, 29, 135
  preoccupancy, 19
  pressure, 172, 186
Cleansers, 141
Clerical office, maintenance, 185
Cloak closet, 109
Closets (see Custodial closet)
Coat hooks, 142
Coils, 175, 176
Collapsible tables, 115
  illustrated, 5
College maintenance facility, 180
Color coding, 103, 152, 170
Color conditioning, 120
Column, 161, 167, 170
  illustrated, 108
Commode-type urinal, 143
Compactor, illustrated, 111
Compression of masonry wall, 49
Concrete, 67, 138
Concrete block, 3, 86, 91, 138, 140, 192
Concrete floor (see Floors, concrete)
Concrete paving, 38
Condensate, 157, 172, 177
Condenser unit, 119
Conductive floor, 63

Conduit, 162
Conference rooms, 181, 182
Consultants, maintenance, 5, 13
Contract cleaning, 26, 29, 135
Contractors, 18, 29
Control board, 182
Control center, 180
Controls, 171, 174, 175, 178
Conveyor systems, 161
Coolant, 172
Cooling tower, 176
Coordinators, 12
Copings, stone, 60
Copper-tubing coils, 175
Cork floors, sound-absorbency of, 69
Corner guards, 39, 95, 128
Corridors, 94, 101, 106, 138, 154, 163, 171, 184, 185
Corrosion, 129, 157, 175, 177, 191
Corrosive vapors, 57
Courtyards, 38
Cove bases, 91, 92, 115, 151
Cranes, 170
Cross-infection control, 27, 65
Cubicle curtains, 117
   illustrated, 108
Curbs, 38, 41, 63, 94, 117, 122, 164, 189
   catwalk, 160
Curtains, 125
   air, illustrated, 48
   cubicle (*see* Cubicle curtains)
Custodial cabinet, 192
   illustrated, 189
Custodial cart, 186
   illustrated, 190
Custodial closet, 10, 128, 132, 146, 156, 185, 187-193
   illustrated, 189, 190
Custodial facilities, 185-187
Custodial maintenance, 25
Custodial manager, 13
Custodial personnel, 182
Custodians, 135

Dado, 88
Data processing, 162, 188, 194
Decorators, 4, 11, 13
Deflector hood, 59
Degreasing, 172

Deodorizer, 163
Design committee, 11, 12
Design file, 13
Desks, 114
Diffusors:
   air, 100, 101
   light, 168
Dimmers, 176
Dispensers, restroom, 9
Displays, 114
   illustrated, 116
Dock bumper, 44
Dock leveler, 44
Docks, 109, 112, 180, 185
Door closers, 129, 130
Door design, 130-133
Door jamb, 90
Door stop, 131
Doors, 24, 41, 107, 117, 127-130, 146, 192, 193
   acoustic, illustrated, 5
   automatic, 129
   entrance, 41
   exterior, 129, 130
   frames for, 90, 127, 132
      illustrated, 129
   glass, 130
   hardware for, 130, 131
   illustrated, 189, 190
   louvered, 132, 192
   numbering, 132
   pulls on, 129
   revolving, 7, 130
   rubber, 132
      illustrated, 133
   trim for, 132
   wood, 132
Dormitories, electric outlets for, 164
Downspouts, 57, 58
Draining, 176, 180, 184
Drains, 22, 43, 45, 63, 109, 137-139, 143, 146, 155, 157, 165, 172, 173, 177
   illustrated, 189
   storm, 24, 40
   trap, 139
Draperies, 103, 106, 125
   illustrated, 126
Drawings, construction, 29
Drinking fountains, 44, 112, 118, 157

Drip-fluid cabinets, illustrated, 137
Drip pans, 157, 172, 177
   illustrated, 189
Dry wall (*see* Gypsum wallboard)
Ductwork, 101, 103, 122, 137, 170, 172, 176
Dust, 18, 19, 162, 164, 177

Ecology, 15
Economic analysis, 13
Egg-crate louvers, 79, 168, 169
   illustrated, 102
Electric enclosures, 164, 165
Electric hand dryer, 150, 163
Electric outlets, 79, 83, 87, 162, 192
   illustrated, 108
Electric panel, 162, 188
Electric receptacle (*see* Electric outlets)
Electric systems, capacity of, 109, 161
Electrical equipment room, 64, 171
Electrical maintenance, 161
Electrostatic filtration, 120, 177
Elevators, 8, 28, 46, 77–80, 91, 109, 162, 188
   control room, 80, 132, 188
   freight, 80
   lobby, 162
   shaft, 80
Embossed tile, 3
Employee handbook, 32
Engineers, 4, 5
Entrance areas, 28, 99, 128
Entrance doors, 41
Epoxy resin, 82, 85, 88, 91, 93 138, 140, 192
Epoxy terrazzo, illustrated, 5
Equipment record, 171
Erosion, 34
Exhaust, smokestack, 173
Exhaust fan, 119, 137, 138, 165, 177
Expansion joints, 64, 152, 161
Explosive vapors, 63
Exterior lighting, 167
Exterior surfaces, 98

Fabrics, 113, 174
Facilities engineers, 13

Factory (*see* Industry)
False ceiling, 27
Faucets, 142, 146, 155, 156, 191
   bibb-type (*see* Bibb-type faucet)
Fences, 44
Fiber glass, 56, 118, 123, 140, 141, 147
File cabinets, 117
Filters, air conditioning, 22, 27, 172, 174, 177
Fin-tube heating, 178
Fire alarms, 31, 157
Fire code, 184, 188
Fire extinguishers, 95
Fire prevention, 104, 160
Fire protection, 110, 127, 176
First aid room, 182
Fixtures, 117–119
   for restrooms, 9, 10, 142–147, 153, 157
   standardization of, 153
Flag raising, 167
Flammables, 183
   illustrated, 184
Flashing, 58
Floodlights, 167
Floors, 62, 161
   acrylic/wood, 70
   asphalt tile, 67
   carpet, 2, 7, 8, 20, 28, 46, 47, 64–66, 78, 81, 93, 103, 106, 128, 138, 157, 193
   ceramic tile (*see* Ceramic tile)
   coating for, 28
   concrete, 67–69, 138, 192
      illustrated, 189, 190
   conductive, 63
   cork, 69
   drains, 22, 25, 31
   grout, 9
   linoleum, 71
   machine, 186
      illustrated, 189, 190
   marble, 71
   mat, 7
   mosaic tile, 71
      illustrated, 72
   pedestal, 63, 74, 159
   plastic laminates, 71
      illustrated, 73
   plastic tile, 73

Floors (*Cont.*):
  resilient, 20, 78, 80, 138, 157
    illustrated, 139, 190
  rubber tile, 74
  stone, 74
  terrazzo, 75, 138, 192
    illustrated, 190
  tile replacement, 62
  travertine, 76
  valves, 144
  vinyl, 70, 76
  vinyl asbestos, 8, 25, 76, 80
    illustrated, 77
  wood, 77, 157
    impregnated, 70
Fluorescent lights, 163, 167, 168
  illustrated, 169
Fluorescent tube crusher, 167, 187
Flush tank, 143
Flush valve, 143, 144, 156
Flusher, 143
Food processing, 88, 98, 100, 118, 119, 128, 134, 155, 168
Fork trucks, 132, 164
  illustrated, 133
Formica, General Electric's, 63, 92, 131
Foundations, 11
Foyers, 99
Freight elevators, 80
Furniture, 28, 88, 112–117, 162, 193
  built-in, 112
  moving, 32, 95

Gallery, service, 160
Galvanized metal, 155, 158, 175
Garages, 179, 180, 183
Gardens, 36
Germs, 27, 65, 138
Glass, 49, 121, 122, 130, 132, 147
Glass block, 122
Glazed tile, 88, 91, 93
Grab bars, 141
Graffiti, 49, 82, 85, 93, 101, 112, 116, 134, 136, 138, 141, 147, 163
Granite for wall surfaces, 93
Grating, 6, 44, 45, 130, 155, 160
  illustrated, 46
Grid carpet, 6

Grills, 116, 137, 138, 192
Grinding, 19, 22
Ground cover, 34
Ground fill, 22
Grounds, 24
  slope of, 34
Grounds-keeping closet, 41
Grout, 9, 31, 67, 71, 88, 91, 138, 140
  illustrated, 75
Guard (device), 176
Guardrail, 94, 117, 154
Gutters, 57
Gymnasiums, wall coverings for, 90
Gypsum wallboard, 89, 91, 103, 142

Hand cleaners, 148
Hand dryers, electric, 150, 163
Handrails, 51, 79, 82
  illustrated, 96
Hardboard, illustrated, 89
Hardware, 130, 131
Hazardous storage, 183
  illustrated, 184
Headboard, 116
Heat (during construction), 20
Heat control, 119, 120, 123
Heating, 100, 127, 174, 177–178
  steam, 177
  strip, 178
Hinges, 116
Hoists, 161, 171, 187
Homogeneous vinyl for flooring, 70
Hospitals, 26, 65, 88, 100, 107, 117, 128, 142, 148, 194
  illustrated, 96
Hot-water system, 177
Hotels, electronic room-notification systems for, 107
Humidity, 162, 180
Hydrants, 156

Ice-melting system, 7
Identification of mechanical and electrical equipment, 165, 171, 176
Illumination (*see* Lighting')
Impregnated wood for floors, 70
Industry, 145, 156, 160, 162, 165, 168

Inflation, effect of, 11
Innovations, testing, 18
Inspection, 160, 161, 174
Inspectors, full-time, for surveillance during construction, 17
Insulation, 60, 122, 154, 158, 165
Integrated ceiling, 100

Janitor closet (*see* Custodial closet)

Key cabinet, 132
Key system, 132
Kick plates, 131
Kitchens, 98, 156

Laboratories, film-processing, lighting for, 167
Ladders, 57, 156, 161, 173, 184, 186
Lamp bases, 117
Lanai, 45
Landscaping, 33–38, 183
Latex paint, 87
Laundry chutes, 157
Laundry machine, 118, 187
Laundry rooms, 156
Lavatories, 142, 148, 157
Lawn sprinkler, 31, 34
Lawns, 33
Lay-in ceiling, 98
Life expectancy as factor in economic analysis, 14
Light globes, 168
Lighting, 27, 83, 85, 87, 93, 100, 102, 104, 138, 163, 165–169, 171, 192
 exterior, 41
 fixtures, 166, 168
 illustrated, 189, 190
Linoleum, 71
Lintels, 50
Litter, 134, 136, 147, 151, 163
Loading dock (*see* Docks)
Lobbies, 7, 134, 162
Locker rooms, 100, 134, 135, 163, 180, 182
Lockers, 119
 illustrated, 151

Lounges, 134, 167, 181, 182
Louver-cleaning machine, 169
Louvered doors, 132, 192
Louvers:
 egg-crate (*see* Egg-crate louvers)
 exterior, 127, 176
Lubrication, 160, 171, 172

Machine shop, 180
Magnetic board, 181, 182
Maintainability, definition of, 1
Maintenance areas, 163
Maintenance building, 179
Maintenance closet, 185
Maintenance consultant, 5, 13
Maintenance control center, 180
Maintenance facilities, 179–194
Maintenance instructions, 29
Maintenance manager, 5, 13, 17
Maintenance offices, 180, 181
Maintenance operations manual, 30
Management, 12
Manholes, 152
Manual, maintenance operations, 30
Marble, 49, 71, 76, 82, 93
 illustrated, 50
Masonry, 50, 94
Masonry wall, compression of, 49
Mastic application, 8, 20
Matting, 45, 46
Mechanical equipment, maintenance of, 169–174
Mechanical equipment room, 132, 155, 160, 171
Mechanical floor mats, 7
Mental institutions, maintenance facilities in, 136, 180
Micarta, Westinghouse's, 63, 92, 131
Microfilm record of construction drawings, 29
Mirrors, 147
 illustrated, 150
Model scale, 169
Modesty strips in restrooms, 142
Molding, 162
Mop hanger, 192
Mopping outfit, illustrated, 190
Mosaic tile, 71
 illustrated, 72

Index

Moving equipment, furniture, and supplies, 32
   specialized equipment for, 32
Mowers, 35, 37
Mowing strip, 35
Mullions, 130
Muriatic acid, 25
Music system, 162

National Electric Code, 164
National Environmental Balancing Bureau (NEBB), 31
National Sanitation Foundation, 119
NEBB (National Environmental Balancing Bureau), 31
Noise in central vending room, minimizing, 112
Nursing homes, 26, 107, 117, 164

Odor control, 65, 134, 137, 138, 143, 152, 163, 174
Office equipment, 162
Office landscaping, 94, 119, 127
Offices, 162, 180, 188
Operating suite, 47, 194
Organization chart, 181

Paint, 87, 91, 93, 103, 138, 140, 158, 192
   illustrated, 2, 190
Painting, 50, 68, 84, 89, 97, 101, 116, 133, 137, 154, 160, 161, 169, 173, 178
Paneling, 88
Panels, access (*see* Access panels)
Parapet, 60
Parking, 39, 155
Parquet floors, 77
Partitions, 94, 96, 121, 143, 146, 159, 163
   restroom, 9, 10, 142
Parts stock, 171
Paving, 24, 35, 38–41
Pea gravel, 41, 45
Pedestal floor, 63, 74, 159
Pegboard, illustrated, 189
Personnel carrier, 164, 183, 186
Pest control, 148, 161

Pharmaceuticals, 88
Pilaster, 93
Pilferage, 133
Pilot plant, 162
Pipe sleeve, 157
Pipe supports, 158
Piping, 34, 101, 103, 104, 122, 139, 143, 152, 157, 161, 164, 170, 172
Pitch pocket, 59
Pivoting windows, 120
Plans (construction drawings), 29
Plant (*see* Industry)
Plantings, 24, 36, 106, 183, 185
Plaster, 82, 90, 103
Plastic laminate, 63, 71, 78, 79, 89, 92, 93, 114, 131, 140
   illustrated, 115
Plastic tile, 73
Plastic window panes, 122
Plumbing, 118, 152, 188, 191
Plumbing chases, 153
Polyester, 140
   illustrated, 146
Porcelain enamel, 90, 169
   illustrated, 90
Posters, 93
Power plant, 179
Power sweeper, 186
Prefabricated walls, 94
Preoccupancy cleaning, 19
Pressure cleaning, 172, 186
Preventive maintenance, 14
Production area, 156
Proximity flusher, 143
Public address system, 162
Public schools (*see* Schools)
Pumps, 157, 173
Push plates, 131

Quarry tile, 68
Quickborner system, 106

Racks, storage, 186
Radiant heating, 178
Radiator cover, 119
Radio-location system, 182
Railings, 161

Ramps, 41, 43, 83
  illustrated, 40
Receptors, 139, 172, 186, 188
  illustrated, 189, 191
Red label on flammable materials, 183
Refinishing areas, 185
Reflector (light), 169
Refrigeration, 118
Relamping, 161, 165
Renovation, 20
Repair areas, 185
Research facility, 162, 180
Resilient floor (see Floors, resilient)
Restaurants, lighting for,·167
Restrooms, 9, 98, 112, 127, 134–152, 155, 163, 180, 182, 188
  equipment for, 147–152
  fixtures for (see Fixtures, for restrooms)
  floors in, 138
  layout of, 135
  partitions, 9, 10, 142
  stalls, 138, 140, 141
  surfaces in, 138–142
  walls in, 140
Retaining walls, 34
Revolving doors, 7, 130
Roof bond, 54
Roof drains, 25, 31, 57
Roof hatches, 188
Roofs, 53–60
Room-notification system, electronic, 107
Rubber tile, 74
Runners, 47

Safety hazards, 164, 172
Safety showers, 155
Safety valves, boiler, 173
Sand finish, 103
Sandblasting, 25
Sanitary napkin dispenser, 147
Sanitation, 134, 137
Scaffolding, 187
Schools, 107, 136, 178, 180
Scotchgard, 113
Screens, 116
Scrubbers (see Automatic scrubbers)

Sealant, glass, 124
Seat covers, toilet, 144
Seating, locker-room, 152
Security, 106, 112, 119, 122, 167, 183
Security office, central, 112
Service area, 163
Service gallery, 160
Sewer gas, 139
Shades, 125
Shaft pit, 80
Sheet metal, 178
Shelving, 185, 191
  illustrated, 189, 190
Shock, mechanical, 165, 166
Shops, maintenance, 180
Shower curtains, 146
Shower doors, 146
Shower heads, 136
Shower partitions, 146
Showers, 98, 135, 146, 148, 155, 182
Shrubs, 24, 37
Sidewalks, 35
Siding, illustrated, 51, 52
Signs, 27, 35, 51, 93, 147, 165, 171
Siphon breaker, 146, 156, 191
Skylights, 104
Slate, 81
Slipping, preventing, 67, 69, 172
Smoke detector, 31
Smokestack, 173
Smoking, damage caused by, 114, 174
Smoking regulations, 107
Snow melting, 40
Soap, hand, 148
Soap dispenser, 148
  illustrated, 149
Soil compaction, 23, 24
Soil entrapment, 44
Solution tank, 10, 169, 186, 191
Sound control, 119, 127
Specifications, construction, 19
Splash area, 41
Splash block, 57
Sponge mat, 47
Sports arenas, restroom facilities in, 135
Spray cleaning, 136, 157
Sprinkler systems, 31, 41, 155, 156
  stoppers for, 157

Index

Stainless steel, 49, 59, 78, 82, 118, 124, 130, 137, 147, 151, 173, 175, 191, 192
  illustrated, 146
Stains, 31, 62, 134, 138, 143, 174
Stair landings, 8, 45, 46, 83, 163, 188
Stair nosings, 51
Stair treads, 50, 81, 82
Stairs, 80–83, 183
  outside, 50
Stairwells, 28, 98, 166
Standardization:
  of maintenance planning, 18
  of plumbing fixtures, 153
Steam, 109, 172
Steam cleaning, 10, 155, 161, 172, 186
Steam heating, 177
Steam piping, 157
Steam trap, 177
Stiles on aluminum doors, side, 132
Stone, 49, 74, 81, 89, 93
  illustrated, 50
Storage areas, 10, 106, 128, 162, 163, 171, 179, 183, 184
Storage cabinets, 186
Storage racks, 185
Storm drains, 24, 40
Strainers, 156, 172
Strip heating, 178
Structural design, 159–161
Sump pan, 175
Sun screen, 126
Superclean rooms (white rooms), 47, 194
Supervisors, maintenance, 181, 182
Suspended ceiling, 98
Switches:
  electric, 138, 164, 167
  panel, 164

Tables, collapsible, 115
  illustrated, 5
Tacky mats, 47
Telephone equipment room, 132, 188
Telephones, 101, 112, 162
Termite control, 33
Terracing, 34

Terrazzo, 22, 24, 25, 63, 75, 80, 128, 138, 145
  illustrated, 146
Testing and balancing of air and hydronics systems, 31
Theaters, lighting for, 167
Thermostats, 175, 176
Thresholds, 128, 138
Tile, glazed, 88, 91, 93
  (*See also* Ceramic tile)
Time standards, 30
Timeclocks, 117, 182
Timed flushers, 143
Tissue dispensers, 141
Toilet rooms (*see* Restrooms)
Toilet stalls (*see* Restrooms, stalls)
Toilets (*see* Water closet)
Tool holder, illustrated, 189
Tools, 171
Topsoil, 22
Towel dispensers, 147, 149, 151
  illustrated, 150
Trade shops, 180
Trailer parking, 39
Training areas, 10, 181
Transformer vaults, 132
Transformers, 119, 164
Transportation terminals, restroom facilities in, 135, 142
Trap (*see* Drains, trap)
Traverse rods, 125
Travertine, 76
Trees, 24, 35, 37
Troweling, 20
Truck dock, 43
Trucks, 132, 133, 164
Tube crusher, 167, 187
Tubs, 141, 146, 148
Tunnels, 160
Turf substitute, 35

Unions, pipe, 156
University maintenance facility, 180
Urinals, 10, 22, 143, 155
  illustrated, 144
  screens for, 143
    illustrated, 144
Utility outlets, 174
Utility sinks, 146, 172, 188, 191

Utility sinks (*Cont.*):
 illustrated, 189, 190
 (*See also* Receptors)

Vacuum (machine), 186, 187, 191, 193, 194
 illustrated, 189, 190
Vacuum cleaning, 161, 172
Valences, 11, 120
Valve boxes, 35
Valves, 24, 35, 144, 152, 153, 155, 156, 158, 170, 172, 175, 177
Vandalism, 53, 57, 101, 112, 116, 119, 121, 133, 134, 136, 141, 143, 147, 160, 161, 163, 166, 167
Vaults, electric, 165
Vehicles, 94, 117, 122, 132, 164, 180, 183
 illustrated, 133
Vending machines, 111, 117, 118
Venetian blinds, 11, 124
Ventilation, 9, 100, 119, 122, 126, 137, 164, 174, 180, 192
Vermin (*see* Pest control)
Vestibules, 7, 128
Vibration, 165, 166
Vinyl asbestos (*see* Floors, vinyl asbestos)
Vinyl floor (*see* Floors, vinyl)
Vinyl wall covering, 82, 84

Wage rates, 14
Walkways, 7, 24, 38, 39, 45, 112
 illustrated, 37
Wall hangings, 87
Wall washing machine, 94
Wallboard, 89, 91, 103
Wallpaper, 85
Walls, 24, 88–93, 112, 117, 136, 140, 152, 157, 161, 166, 176, 192, 193
 coverings for, 82, 84–87, 93
  illustrated, 2
 design for, 93–95
 illustrated, 190

Washbasin, circular, 136, 145
 illustrated, 146
Washing machines, 10
 wall, 94
Washrooms (*see* Restrooms)
Waste chute, 157
Waste compactor, 183
Waste container, 41
 illustrated, 190
Waste disposal 41, 43, 109, 167
 illustrated, 110
Waste receptacle, 78, 79, 83, 110, 143, 147, 151
 illustrated, 108, 148, 150
Waste removal, 19, 30, 111
Water closet, 143–145, 155
Water closet seat, 144, 145
 self-raising, illustrated, 145
Water cooler, 117
Water fountain, 157
Water main, 155
Water outlets, 109, 173
Water-softening equipment, 157
Waterproofing, 164
Welding, 161
Wet cleaning, 28
Wet vacuums (*see* Vacuum)
Wheelchair ramps, 41, 43
Wheelchairs, 141
 illustrated, 96
Wheels, 117
White room, 47, 194
Window cleaner, automatic, 124
 illustrated, 125
Window ledge, 122
 illustrated, 123
Windows, 10, 90, 119–127, 176
 pivoting, 120
 washing, 120, 127
Windowsills, 119
Wiring, 106, 118, 182, 192
Wiring raceway, 101
Wood flooring, 77, 157
 impregnated, 70
Wood paneling, 88
Work tables, 186
Wright, Frank Lloyd, 121